T0073590

THE TRILOBITE
COLLECTOR'S GUIDE

THE TRILOBITE COLLECTOR'S GUIDE

ANDY SECHER

FOREWORDS BY
RICHARD FORTEY
AND MELANIE J. HOPKINS

Columbia University Press   New York

Columbia University Press
Publishers Since 1893
New York   Chichester, West Sussex
cup.columbia.edu

Copyright © 2024 Andy Secher
All rights reserved

Library of Congress Cataloging-in-Publication Data
Names: Secher, Andy, author.
Title: The trilobite collector's guide / Andy Secher.
Description: New York : Columbia University Press, [2023] | Includes index.
Identifiers: LCCN 2023027875 (print) | LCCN 2023027876 (ebook) |
    ISBN 9780231213806 (hardback) | ISBN 9780231560047 (ebook)
Subjects: LCSH: Trilobites—Collection and preservation.
Classification: LCC QE821 .S434 2023 (print) | LCC QE821 (ebook) |
    DDC 565/.39—dc23/eng/20230614
LC record available at https://lccn.loc.gov/2023027875
LC ebook record available at https://lccn.loc.gov/2023027876

Printed in the United States of America

Cover design: Milenda Lee

COVER IMAGE: *FOULONIA PEREGRINA* (DEAN, 1966)
Lower Ordovician, Floian Stage; Upper Fezouata Formation; Zagora area, Morocco; largest specimen 6 cm

COVER IMAGE (*FRONT FLAP***)** *CONOMICMACCA ALTA* (LINAN AND GONZALO, 1986)
Lower Cambrian, Bilbilian; Láncara Formation; Cantabrian Mountains, León Province, Spain; each trilobite 3.2 cm

COVER IMAGE (*BLACK FLAP***)** *RADIASPIS SP.*
Middle Devonian, Eifelian; El Otfal Formation; Jebel Ofaténe, Morocco; 6.6 cm

TITLE PAGE IMAGE: *GABRICERAURUS DENTATUS* (RAYMOND AND BARTON, 1913)
Upper Ordovician; Bobcaygeon Formation; Deseronto, Ontario, Canada; 10.8 cm

This book is dedicated to hard-core trilobite collectors everywhere, an admittedly quirky group (of which I'm proudly a member) comprised of people who would much prefer to spend their Saturday afternoons searching for creatures that expired more than half-a-billion years ago than mowing the lawn or traipsing through the local mall. What motivates these folks to consider the acquisition of ancient invertebrates—whether in the field, on the internet, or at major rock and mineral shows—is certainly open for discussion and debate. Most people drawn to these incredible arthropods state that their fossiliferous fascination stems from a healthy combination of curiosity and compulsion. Those two factors together fill their minds with wondrous images of a long gone age and too often empty their wallets as they tenaciously pursue these beguiling relics of the Paleozoic past. It is for these stalwart Deep Time devotees that *The Trilobite Collector's Guide* has been assembled with as much thought, insight, and care as I could consistently muster.

Contents

(OPPOSITE PAGE) ***DIONIDE MAREKI* HENRY AND ROMANO 1978**
Upper Ordovician, Llandeilo series; Valongo Formation; Covelo, Portugal; 9 cm

BEETLES OF THE PALEOZOIC

Dinosaurs are familiar to every child and to most adults as the embodiment of exotic and thrilling extinct animals. Their complex names (think of *Pachycephalosaurus*) trip off the tongues of five-year-olds even before they can read, and one could be forgiven for thinking that the geological past was populated by no other creatures at all. Yet long before the dinosaurs had evolved the seas thronged with animals just as bizarre and charismatic. First among these were the trilobites. Like dinosaurs, they are extinct; but unlike the giant reptiles, they left no descendants. They attest to vanished worlds thronging with strange life-forms.

Trilobites include many spectacular fossils, some as large as lobsters, others smaller than peas; many of them are decked with arrays of spines that would put a porcupine to shame, others are smooth and polished. There are many thousands of known species—so many that they have been dubbed "the beetles of the Paleozoic." Some were equipped with spectacular eyes carrying giant lenses that are still clearly preserved. It is easy to imagine them peering through the seas in which they thronged in search of prey or predator. Others lost their eyes altogether and grubbed around on deep seafloors beyond the reach of light. Many could roll up into tight balls when threatened. Some were streamlined swimmers in the open seas, and many more lived on or near the seafloor, enjoying the full span of lifestyles displayed by present-day crustaceans (crabs, lobsters, shrimps, barnacles, and many more) ranging from predators to grazers and filter feeders. Trilobites evolved into such remarkable shapes that they are unmatched in today's oceans—consider the species of *Walliserops* that carried a massive trident on the front of their head. Surely no dinosaur was as extraordinary!

I spent my working life as a scientist describing new trilobites from my desk in the Natural History Museum, London, totting up almost 200 species "new to science," as the jargon has it, during more than 50 years in the fossil trade. Many of these I collected myself in the field, surely the most enjoyable part of an enviable job. However, most of my scientific results were published in journals unavailable to many naturalists. Because trilobite "shells" fell into fragments after the death of the animal, many of the species I named were based on incomplete material, particularly heads and tails (cephalon and pygidium, formally speaking).

To find an articulated, whole trilobite was a comparatively rare thrill in many localities, and we did not have days to spend looking for the perfect complete example. But they were out there to be discovered with patience, determination, and not a little luck.

This book is a treasure house of the best of the best, the fruits of Andy Secher's incomparable collection from all over the world. Wherever that perfect trilobite could be found illuminating their 250-million-year history, a trilobite enthusiast was not far behind the first discovery. Too often these specimens were secreted in private collections, but Andy wants everyone to enjoy the beauty and character of his favorite fossils. When I wrote my own book more than 20 years ago (*Trilobite! Eyewitness to Evolution*), I illustrated a good number of fine specimens but did not have the benefit of modern color reproduction, and my intention was primarily to explain the science and stories that made up trilobite lore. This new book is a celebration of the whole gamut of trilobites using the best specimens and illustrations possible. It can be enjoyed first as a picture book by anyone who wishes to see the astonishing variety of forms achieved by these animals through eras when the face of the planet was utterly different from the present day.

Trilobite fossils occur in sedimentary rocks on every continent—even in Antarctica—and some localities have become justly famous for the profusion of fossils they yield. Of these, some were "worked out" in the last century, at a time when trilobites were crucial to unraveling the narrative of geological time. Rocks of different ages had different suites of trilobite species, and pioneers such as Charles Dolittle Walcott used them to align rocks of the same age on different continents—they were then "cutting edge." In recent years, many new trilobite-bearing localities have been discovered, most notably in Morocco and China. Particularly in Morocco the desert outcrops allow for prodigious feats of collecting, and the preservation is just right to extract the finest details from their rocky burial grounds. This book is an around the world tour through some of the most famous trilobite sites. Some of the most precious and remarkable beasts are beautifully illustrated to underpin the importance of these localities not just to science but to inspiration. Each trilobite has a story to tell, and each locality is a sum of those stories that provide us with a history. Trilobites also can be enjoyed just for their own beauty or unique personality—or just plain weirdness. Who knows? Maybe one day the name *Phacops* will be as familiar to us as that of *Tyrannosaurus*.

Richard Fortey, FRS
Former head of fossil arthropod research
Natural History Museum, London

Foreword

10 REASONS TO READ THIS BOOK

Andy Secher has the most impressive private trilobite collection I've ever seen, unmatched in terms of size and breadth. To be a truly great collector, however, is also to be an ambassador. In this book, you will see many stunning photographs of Andy's collection, and you will be infected by Andy's enthusiasm not only for the specimens themselves but for sharing them and their significance with the world. Something can be learned from each and every specimen in this book, and Andy has skillfully arranged everything in top 10 lists. You can let him guide you from start to finish, or you can jump around, reading in any order of interest to you. Always wondered how big trilobites could get? Check out "10 Beyond Big Trilobites." Getting ready to go to the Natural History Museum in Prague? Flip to "10 Beautiful Bohemian Trilobites."

In the spirit of the highly enjoyable book Andy has produced, I offer the following list.

TOP 10 REASONS TO READ THIS BOOK

To be impressed by how diverse life has been on Earth.
Some scientists estimate that more than 99 percent of all species that have lived on Earth are now extinct. Without the fossil record, we wouldn't know that more than 22,000 species of trilobites existed (and counting). That's a lot of trilobite spines (10 Spectacularly Spined Trilobites) and a lot of trilobite eyes (10 Engagingly Eyed Trilobites).

To be dumbfounded by how old trilobites are.

Trilobites are one of the first macroscopic animal groups to appear in the fossil record (10 Outrageously Old Trilobites), and they were around for 270 million years before going extinct (10 Last-in-Line Trilobites). Plenty can—and did—happen in that amount of time (10 Time and Trilobite-Related Theories).

To be amazed by the different ways trilobites are preserved.

Trilobite fossils are (mostly) just the "hard parts"; that is, the biomineralized exoskeleton of the trilobite animal. However, there is a lot of variation in how the exoskeleton was preserved (10 Cleverly Colored Trilobites), and even some special cases in which the "soft parts" such as legs and gills have also been preserved (10 Significant Soft-Tissue Trilobites).

To be reminded that what is now rock was once living and breathing.

Because trilobite fossils are mineralized, it is easy to forget that those specimens were once animals crawling around the ocean floor, digging in the muck, trying to evade danger (10 Rock 'n' Roller Trilobites) but not always succeeding (10 Precariously Predated Trilobites), and just living their trilobite lives (10 Tales Told by Trilobites).

To be surprised by how common some trilobites are.

Some trilobite specimens are valued by collectors and museums because they are unusual or rare (10 Ridiculously Rare Trilobites). Those specimens might be rare because they belong to species that were rare when alive, because they were rarely preserved, or because the rocks they were preserved in are rarely exposed. But there are many others that are ridiculously common (10 Most Common Trilobites).

To be carried on a trip around the world.

Trilobites are found everywhere in the world: China, Australia, Canada, the United States, Sweden, England, and Morocco, just to name a few of the countries featured in this book. After all, trilobites knew no political or geographic boundaries (it's the type and age of the rock that matters).

To be inducted into the cult of location.

There's a reason that so much of this book focuses on localities: like real estate, it's all about location, location, location. Where a trilobite specimen is found provides crucial information about how old it is and what the paleoenvironment was like. It's also information that could lead to additional discoveries in the future (10 Obscure [But Still Significant] Trilobite Localities).

To be briefed on the how-to of collecting (10 Trilobite Collecting Tips), preparation (10 Pertinent Preparation Steps), and curation (10 Key Curation Details).
It turns out that a lot of steps are needed to go from discovery to publication-quality photograph!

To be introduced to past and present trilobite enthusiasts.
Sprinkled throughout almost every list in this book are personalities—those who made the discoveries, those who kept going back, and those who tracked down lost localities. Trilobites have been capturing people's imaginations for at least 300 years (10 Essential Figures in Trilobite History) and possibly much longer.

To be inspired by the fact that paleontology is for everyone.
Although some special sites are protected for their historical or scientific importance, many publicly available roadcuts and quarries, river beds and cliffsides, and desert outcrops contain trilobites waiting to be discovered by anyone. There are also many ways—museum visits (10 Top Trilobite Museums), fossil shows (10 World-Class Fossil Shows), and books (10 Must-Read Trilobite Books)—to enjoy trilobites without getting dirty. Whatever way is best for you, Andy's book will inspire you to get out there and start collecting!

Melanie J. Hopkins
Chair, Division of Paleontology
Curator-in-Charge, Fossil Invertebrates
American Museum of Natural History, New York

PALEOZOIC

541–252 MILLION YEARS AGO

NUMBER OF EXISTING TRILOBITE ORDERS

MILLIONS OF YEARS

Orders	Period	Duration
1	**Permian** 299–252 Million Years Ago	52
1	Pennsylvanian	19
1	**Carboniferous** 359–299 Million Years Ago Mississippian	41
6	**Devonian** 419–359 Million Years Ago	57
7	**Silurian** 444–419 Million Years Ago	28
9	**Ordovician** 485–444 Million Years Ago	41
9	**Cambrian** 541–485 Million Years Ago	54

Time chart designed by Mark Ault

THE TRILOBITE
COLLECTOR'S GUIDE

INTRODUCTION

THE TRILOBITE COLLECTOR

It was no less a figure than the former British prime minister Winston Churchill who once memorably remarked, "The farther backward you can look, the farther forward you can see." If such is the case—and who are we to doubt the ever-clever Sir Winston?—then those who collect trilobites must certainly rank among the planet's great visionaries. After all, it is they who routinely look back some half-a-billion years in their appreciation of these remarkable but long gone undersea organisms. And it is they who then do their best to disseminate both their trilobite-derived knowledge and their passion to a less than anxiously awaiting world.

The dedication of these arthropod-obsessed enthusiasts is at least somewhat understandable; trilobites were among the most successful life-forms ever to exist on Earth. Their march through geologic time began in the Lower Cambrian, some 521 million years ago, and lasted until their demise at the end of the Permian, more than a quarter-of-a-billion years later. During that unfathomable expanse of planetary history (which ostensibly covers the entire Paleozoic Era), these highly versatile, hard-shelled animals evolved to fill virtually every available marine niche while producing more than 25,000 scientifically recognized species—including some of the strangest yet strangely captivating creatures ever to crawl through the primeval seas. This almost inexplicable allure, due in no small part to the amazing diversity of morphological design displayed by such tongue-twisting species as

Athabaskia wasatchensis, *Stelckaspis perplexa*, *Hypodicranotus striatulus*, and *Allolichas halli*, has helped transform key members of the trilobite class into some of the most sought-after collectibles in the entire natural history realm.

Each year, thousands of fossil-loving folks from London to Los Angeles travel to rock and mineral shows, peruse the internet, or journey into the field in pursuit of these long-departed ocean-dwelling relics. Many Deep Time devotees are then willing to spend what may initially seem excessive sums in their efforts to procure new species for their ever-expanding assemblage of trilobite treasures. In the last few decades, this nearly fanatical fervor has become increasingly accepted in polite society. Despite their great age, fossilized fragility, and often diminutive size, trilobites now rank second only to the hallowed dinosaur in terms of both their paleontological recognition and their renown. In the years ahead, their popularity, collectability, and sphere of intellectual influence seem destined to increase; after all, it's a lot easier to precisely position a complete, 4-centimeter *Greenops boothi* in your carefully curated cabinet of curiosities than a complete *T. rex* skull!

This brings us to the book you're now holding, *The Trilobite Collector's Guide*, which was expressly constructed to provide a unique window into the long-ago lives of these ubiquitous maritime invertebrates. In all honesty, when my previous Paleozoic-themed volume *Travels with Trilobites* emerged in the spring of 2022, little did anyone (including me) foresee a reason for a follow-up. Heck, what more could possibly be said or seen regarding the World's Favorite Fossil Arthropod? When the unexpected but welcomed call came to gather enough pertinent material to produce a new manuscript of trilobite-infused text, the cephalon scratching began. Hmmm. What to do, and where to go next? A daunting but quickly accepted challenge presented itself. The key question became how to best present simultaneously provocative and compelling content that was as fresh, informative, and illuminating as that featured in this book's notable predecessor. At first, the editorial options appeared rather finite. Certainly there were plenty of potential photos to fill any subsequent effort, especially with ready access to one of North America's most eminent private trilobite collections—featuring more than 4,000 choice specimens—only a short stone's throw away. But as additional thought began to be cast on finding a proper thematic direction for this volume, the available avenues slowly began to reveal themselves to be both intriguing and slightly intimidating.

Even in the relatively brief span separating the appearance of these two fossil-fueled books—a veritable geologic nanosecond—myriad previously unknown trilobite species were uncovered in a sequence of hitherto uncharted Paleozoic horizons found across the face of the planet (including sites in Morocco, Spain, Siberia, China, Australia, Greenland, and the United States). These new species have supplied more than enough material to figuratively overflow these pages. When the inspiration supplied by these recently emerging, globe-encircling, paleo-centric outcrops was combined with the perspiration generated by some quasi-clever editorial decisions—including the notion of shining a brief but revealing spotlight on trilobite-bearing localities only casually referenced if not totally ignored previously—the focus of *The Trilobite Collector's Guide* began to gain greater clarity. Even the book's slightly sardonic (and a tad too lengthy) original subtitle, *Everything You Didn't Know You Wanted to Know About the World's Favorite Fossil Arthropod*, was chosen to reflect the thinly veiled truth that there is still so much to learn about trilobites—and that most of the planet's population is only now beginning to recognize the time-tested "animal magnetism" displayed by these incredible invertebrates.

ATHABASKIA WASATCHENSIS (RESSER, 1939)
Middle Cambrian; Langston Formation, Spence Shale Member; Bear River, Utah, United States; 6.3 cm

Following a further bit of analysis, research, and preparation, it was determined that *The Trilobite Collector's Guide* would feature a fast-paced series of concise top 10 lists that covered everything from *Celebrated Cambrian Trilobite Localities*, to *Essential Figures in Trilobite History*, to *World-Class Fossil Shows*, to *Beyond Big Trilobites*, to *Ridiculously Rare Trilobites*, to *Ways to Value Your Trilobites*, to *Pertinent Preparation Steps*. It was thought that such a numerically inclined paleo-scape would present an exciting, enlightening, and perhaps even more easily digestible take on all the fossil-saturated proceedings contained within. One trilobite-related topic seemed to inspire the next, until somewhat surprisingly the initial arthropod embracing abstraction had evolved into a page-turning reality featuring more than 50 distinct top 10 categories—along with hundreds of never before seen color photographs illustrating many of the key trilobite specimens discussed in each list.

All things considered, there are a multitude of reasons to cast additional literary light on these most fascinating of early marine inhabitants.

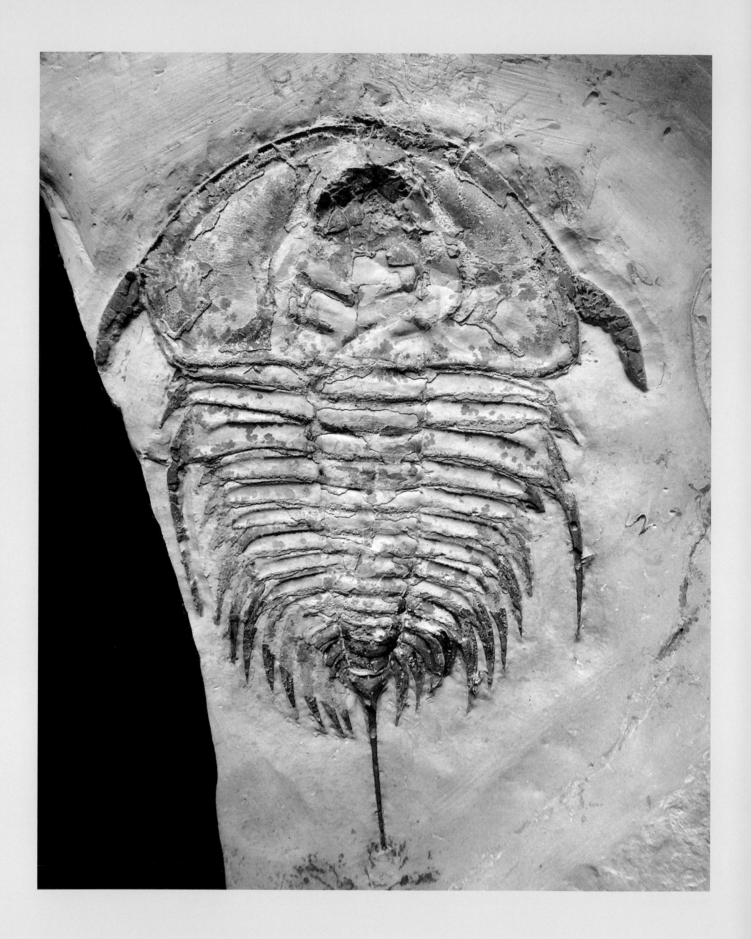

As someone who has long been intimately involved with just about every aspect of the trilobite collecting experience, I believe this *Guide* fills an important and long-standing void in the paleontological field. In stark contrast to the various textbooks, tomes, and treatises that have previously taken direct aim at these ancient arthropods, from day one my primary goal for *The Trilobite Collector's Guide* has been to blend an abundance of information, a degree of illumination, and a healthy dose of fossiliferous fun into a top 10–filled package of pure Paleozoic entertainment.

And to address a frequently asked question, I am certainly not a trained paleontologist, nor do I ever intend or pretend to be one. I am more than content merely being an interested "outsider" (with a few key museum affiliations) who maintains a particular—and some might even say peculiar—fascination for these eminently enchanting fossilized forms. In both style and substance, this guide is written by a collector for my fellow collectors, whether those trilobite enthusiasts happen to reside in Boston, Brisbane, or Barcelona. It's no secret that a hobbyist's prime objective often appears to be in direct contrast to that of a scientist. We crave complete pristine examples, whereas academicians are usually satisfied studying a disarticulated cephalon or a displaced pygidium—as long as those fragmentary trilo-bits help support a pet theory or lend credence to an *au courant* debate. Despite this apparent philosophical disparity, there is no "right way" or "wrong way" to assemble a trilobite collection. Quite simply, you should collect what you like, when you like, how you like. As John Lennon sang so many years ago, whatever gets you through the night!

(OPPOSITE PAGE) **MESONACIS N. SP.**
Lower Cambrian; Combined Metals Member; Pioche Shale, between Klondike Gap and Ruin Wash faunas; Nevada, United States; 9.5 cm

Contained within these glossy, image-packed pages should be just enough sedimentary-steeped substance to satisfy even the most inquisitive academic mind. At the same time, it is hoped that this carefully considered amalgam of trilobite-related words and photos still manages to convey the degree of awe, admiration, and wonder too often missing from more scholarly texts. With all such knowledge now placed in its proper Paleozoic perspective, please find a comfortable seat and let *The Trilobite Collector's Guide* take you on an engaging and enjoyable journey back some 500 million years—to a time when these highly adaptable arthropods were the dominant life-forms in the evolutionarily evocative waters of planet Earth.

THE PHOTOGRAPHS

The nearly 350 color photographs in *The Trilobite Collector's Guide*—all representing specimens drawn from my personal collection unless otherwise noted—have been selected specifically both to display the trilobite class's incredible diversity of form and to reflect the amazing array of genera that appear scattered among this volume's various top 10 lists. From the graceful configurations presented by Cambrian olenellids, to the streamlined shapes exhibited by Ordovician cheirurids, to the ornate intricacies of Silurian lichids, to the magnificent compound eyes that adorn Devonian phacopids, this book's paramount photographic intent is to showcase the impressive morphological heterogeneity that the trilobite line demonstrated throughout its quarter-billion-year trek through evolutionary time—and to do so in as pleasing and accessible a manner as possible. An earnest attempt has also been made to avoid repeating photos or specimens presented in this book's predecessor, *Travels with Trilobites*. More than 90 percent of the images in *The Trilobite Collector's Guide* have never previously been seen—no matter whether

ALLOLICHAS HALLI **(FOERSTE, 1888)**

**Upper Ordovician; Dillsboro Formation; Oldenberg, Indiana,
United States; 6.2 cm**

those observing eyes are of anthropoid, arthropod, or annelid origin. In the pages directly adjoining the various top 10 chapters, key curation details (identifying the photographed trilobite as to its genus, species, author, age, location, formation, and size) have been provided for each featured specimen.

For those wondering, I am responsible for virtually all of the trilobite photos appearing in these pages. I did initially employ a professional photographer to try to capture the essence of these creatures' enigmatic appeal, but I found most of those images to be strangely lacking in what might best be termed "soul." So, for better or worse, I gave it a go myself. I take a degree of perhaps undue pride in my writing abilities, but I fully recognize my limitations behind the lens. Considering some of the memorably madcap photo sessions in which I was involved during my three-decade stint in the rock 'n' roll industry, I found it surprisingly pleasurable to deal with "rock star" subjects that didn't require their agents, managers, or publicists to be perpetually hovering about them during a shoot. Indeed, during the creation of these photographs

the only makeup person required was me, as I gently brushed an occasionally uncooperative bit of dust from a fragile genal spine or a dangerously dangling opisthothorax.

METRIC MEASUREMENTS

Readers will quickly note that all measurements appearing in *The Trilobite Collector's Guide* are presented in metric terms. That's true whether the size in question is indicating the dorsal length of a diminutive trinucleid trilobite or depicting the distance between Paleozoic continents. Most trilobites were relatively small, and measuring these specimens in centimeters rather than inches has become an accepted international standard that I have chosen to impassively embrace. Once that initial metric threshold was breached, switching from feet to meters and miles to kilometers also seemed to make logical sense. These measurements may seem a little cumbersome at first, but for those who need a quick refresher, just remember that 1 inch equals 2.54 centimeters and 1 mile translates into 1.61 kilometers.

(LEFT) GREENOPS CF. BOOTHI (GREEN, 1837)

Middle Devonian; Mahantango Formation, Hamilton Group; Carbon County, Pennsylvania, United States; 4.2 cm

(BOTTOM) PARACERAURUS INGRICUS (SCHMIDT, 1881)

Middle Ordovician, Arenig Series; Volkhov Regional Stage; Putilovo quarry; St.Petersburg region, Russia; 8.1 cm

***BRISTOLIA FRAGILIS* PALMER AND HALLEY, 1979**

Lower Cambrian; Carrara Formation, Thimble Limestone
Member; Emigrant Pass, California, United States; 7.1 cm

CTENOCEPHALUS CF. TERRANOVICUS
(RESSER, 1937)

Middle Cambrian; Chamberlains Brook Formation,
Adeyton Group; Manuels, Newfoundland, Canada; 2.3 cm

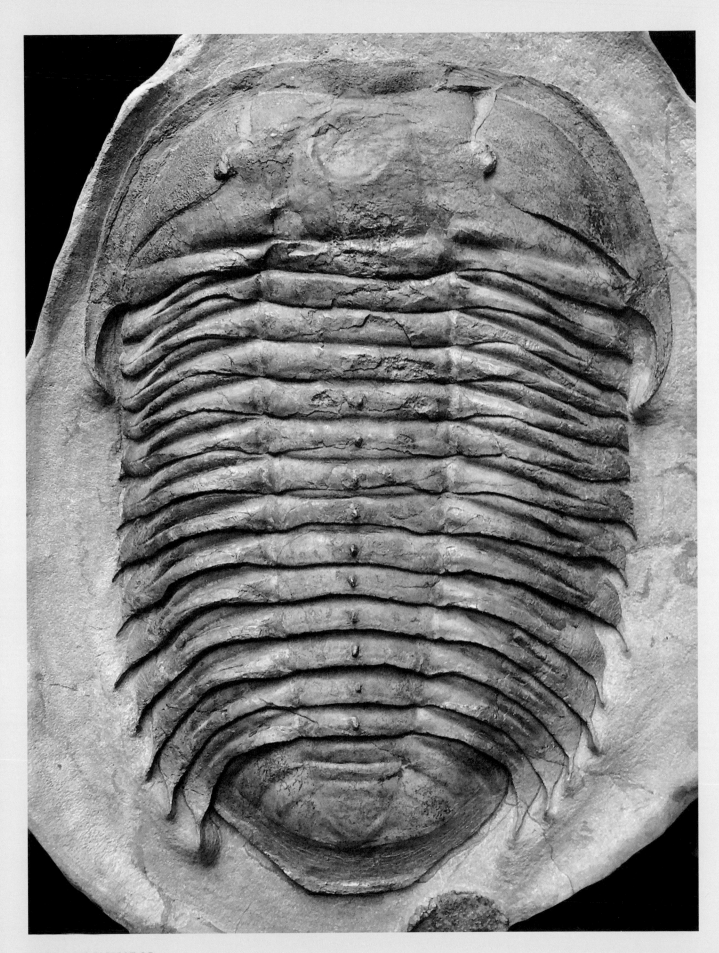

IBEROCORYPHE SP.

Upper Ordovician; Caradoc Series; Agdez, Morocco; 19.3 cm

OLENOIDES SKABELUNDI ROBISON & BABCOCK, 2011

Middle Cambrian; Weeks Formation; House Range, Utah; 16.5 cm

10 CELEBRATED CAMBRIAN TRILOBITE LOCALITIES

541 to 485 Million Years Ago

The dawn of the Cambrian Period corresponded in near perfect synchronicity with the emergence of multicellular life on Earth. In true chicken-or-egg fashion, it is this unprecedented biological blooming that signaled the beginning of the famed Cambrian Explosion. Trilobites served as key index fossils for the entire Paleozoic Era, arising in the Lower Cambrian 521 million years ago, and lasting until the end of the Permian some 270 million years later. Only a few million years after their initial appearance, trilobites were already thriving in those primeval oceans, with thousands of distinct species filling virtually every available ecological niche in the planet's diverse marine habitats. But these amazing arthropods were far from the only creatures living in those early seas. In the Cambrian, a time when oceans dominated the globe and surrounded the southern hemisphere supercontinents Laurentia and Amazonia, trilobite species such as *Bathyuriscus fimbriatus, Wutingaspis tingi, Orygmaspis morrisi*, and *Olenellus cordillerae* were joined by a plethora of other invertebrates. Many of these were nonbiomineralized creatures whose soft-bodied existence has been chronicled in varied locations across the face of the Earth, including paleontological treasure troves in China, Australia, and western Canada. Yet, as dramatic evidence of the destructive powers the planet was going to routinely unleash upon its new inhabitants, after a highly productive 59-million-year journey through evolutionary time, the Cambrian abruptly ended with the advent of Earth's first major extinction event. In the wake of that episode, which may have been triggered by a pronounced drop in marine oxygen levels, 60 percent of

extant ocean life—including many trilobite genera in addition to entire orders of brachiopods, conodonts, and soft-bodied arthropods—forever disappeared from the Paleozoic scene.

The following sections provide an overview of 10 celebrated trilobite-bearing Cambrian locations around the world.

SAKHA REPUBLIC, SIBERIA

Cutting for nearly a thousand kilometers on a south-to-north path through the heart of Siberia's distant Sakha Republic are the powerful Lena and Anabar rivers. On their own merits, these closely aligned bodies of water would most likely retain relative anonymity on the global stage. However, the fact that some of Earth's oldest trilobites have been discovered in the 521-million-year-old sedimentary outcrops carved out by these sister waterways makes them of particular interest to both paleontologists and trilobite enthusiasts around the world. With fossils found within the area's Sinsk and Pestrotsvet formations ranging in size from the diminutive *Delgadella lenaica* (which rarely exceeded 1 centimeter), to the midsized *Bathyuriscellus siniensis*, to the large *Jakutus primigenius* (which often attained lengths of 12 centimeters or more), it is apparent that these deposits were once home to a diverse and advanced fauna—especially considering how early they appear in the trilobite record. Indeed, perhaps the first trilobite in the entire fossil lineage, *Profallotaspis jakutensis*, has been documented in adjacent mudstone layers, marking this remote Siberian outpost as one of unique paleontological significance.

(OPPOSITE PAGE) **BATHYURISCUS FIMBRIATUS ROBISON, 1964**
Middle Cambrian; Wheeler Formation, House Range; Millard County, Utah, United States; 8.2 cm

EMU BAY SHALE, AUSTRALIA

The Emu Bay Shale of Kangaroo Island, South Australia, ranks among the world's most significant Lower Cambrian deposits. Presenting a geologic age and inherent fauna some two million years older than that of the hallowed Burgess Shale of British Columbia, Canada, these 510-million-year-old layers are filled with a varied assortment of Burgess-like soft-bodied creatures as well as with such stratigraphically important trilobite species as *Estaingia bilobata, Balcoracania dailyi, Redlichia takooensis*, and *Redlichia rex*—the latter of which are renowned for emerging as colorfully preserved specimens up to 25 centimeters in length. First noted for its amazing Paleozoic fauna in the mid-1950s, the Emu Bay Shale is believed to have once been part of a semitropical, shallow sea environment in contrast to the more temperate environs proposed for both Burgess and China's similarly primal Chengjiang Biota. Aided by its unusual and rather isolated geographical location in the Cambrian, a few of the Emu Bay's trilobite genera developed remarkable morphological features—with the fossilized remains of *Balcoracania dailyi* displaying more than 90 distinct thoracic segments, the most of any known trilobite species.

WEEKS FORMATION, UTAH

Due to both the unique preservation and stunning diversity exhibited by its more than two dozen species of described trilobites, for many collectors Utah's 500-million-year-old Weeks Formation represents one of the most captivating fossil repositories found in the western United States. Featuring a distinct fauna of generally diminutive trilobites that rarely exceed 4 centimeters in length—including *Tricrepicephalus texanus, Meniscopsia beebei*, and *Norwoodia boninoi*—the Weeks was long thought to represent a well-defined Upper Cambrian bio-zone. However, recent studies—along with the recovery of a rare species of large *Olenoides*—indicate that it is, in fact, one of the last bastions of the Middle Cambrian to be found anywhere in North America. The trilobites discovered here are often preserved in a thick black or brown calcite and are embedded in matrix that can vary from light tan to near-magenta in tone. Such attractive color contrasts have made Weeks material (which also includes the large aglaspid, *Beckwithia typa*, along with an assortment of generally small, soft-bodied organisms) among the most eagerly sought in the world. In addition, following a decade during which the formation's shale layers were commercially extracted for use as garden paving stones, recent legislation has placed much of the area's Paleozoic outcroppings on a "restricted" list, serving to further increase the lure of any and all trilobites that hail from the Weeks.

CHENGJIANG BIOTA, CHINA

In 1984, a thousand years after these fossil-filled deposits were first noted by area residents, Chinese scientists formally described an amazing 515-million-year-old Lower Cambrian fauna located within the Maotianshan Shales of Chengjiang County in China's southern province of Yunnan. Mixed amid a variety of primitive trilobite species, including *Eoredlichia intermedia*, *Yunnanocephalus yunnanensis*, *Wutingaspis tingi*, and *Kuanyangia pustulosa*, were an astounding assortment of soft-bodied creatures, many of which resembled those found in the legendary—although seven million years more recent—Burgess Shale. Material drawn from the various formations that comprise this site (which have become collectively recognized as the Chengjiang Biota) has been known since the tenth century. However, it wasn't until academic expeditions began exploring the

area in the mid-1980s that researchers began to unravel the myriad scientific mysteries that lay within these 50-meter-thick mudstone layers. So far, 185 different species (with over half being arthropods—four being the aforementioned trilobites) have revealed themselves in the Chengjiang Biota, with the possibility that even more are still hidden within these primeval rocks, waiting to be discovered.

PIOCHE SHALE, NEVADA

Just a quick hop from the notorious UFO haven of Area 51, and a mere three-hour drive from the glittering lights of Las Vegas, there has long been a bit of added Paleozoic panache surrounding the Lower Cambrian outcrops that comprise Nevada's Pioche Shale Formation. Although the environs in this part of Lincoln County often present more than their share of challenges to those in search of fossils, especially with summer temperatures that routinely reach 115 degrees Fahrenheit (46 degrees Celsius), the results derived from exploring this fascinating locale are usually worth the inherent difficulties. Captured within these beige-tinged, 514-million-year-old sedimentary layers are key elements representing one of early Earth's most important transitional stages, as the Lower Cambrian fauna slowly began to fade, to be eventually replaced by a flurry of new, Middle Cambrian genera. Featuring an exotic array of closely related trilobite species—including *Bristolia bristolensis*, *Bristolia nevadensis*, *Olenellus fowleri*, and *Olenellus chiefensis*, all of which are preserved within this deposit in a golden-hued hematite—the trilobites of the Pioche Shale comprise some of the most dramatic and aesthetic fossils to be found anywhere in the world. Indeed, when considered along with the other assembled fauna, these trilobites provide an unparalleled view of what a typical reef-filled offshore environment

***BATHYURISCELLUS SINIENSIS* IVANTSOV IN PONOMARENKO, 2005**

Lower Cambrian, Botomian Regional Stage; Sinsk Formation; Botoma River (right tributary of
Lena River), Lena Pillars Park; Southern Sakha Republic (Yakutia), Eastern Siberia, Russia; 7.2 cm

may have been like at this primal stage of life's development on Earth.

BURGESS SHALE, BRITISH COLUMBIA

No other invertebrate locality on the planet enjoys the mystique and mainstream media focus of the Burgess Shale. High in the snow-capped Canadian Rockies lies the Stephen Formation, its sedimentary layers saturated not only with trilobites such as *Olenoides serratus* and *Ogygopsis klotzi* but also with an amazing array of soft-bodied creatures that together provide a special look at a Middle Cambrian realm that existed over half-a-billion years ago. This remote site in British Columbia, first noted in 1886 by railway workers, was subsequently "discovered" by the legendary Charles Walcott during the early years of the twentieth century. Indeed, Walcott's ability to brilliantly promote his Burgess finds through articles in mainstream publications like *National Geographic* helped quickly turn this high-elevation treasure trove into the most recognized invertebrate fossil locale on the planet. Since then the Burgess Shale has remained the subject of intense (and ongoing) scientific research as well as serving as the springboard for dozens of books, internet reports, and TV specials dedicated to trying to decipher the riddles of life on Earth soon after the Cambrian Explosion began. In 1981 UNESCO recognized the importance of the Burgess Shale horizon by declaring it a World Heritage Site, a designation designed to protect this location's amazing fossil fauna for future generations.

PARKER QUARRY, VERMONT

Located in the outskirts of a small town in northern Vermont is a legendary trilobite-bearing site known as the Parker Quarry. As far back as the late nineteenth century, renowned American paleontologists including James Hall and Charles Walcott examined specimens emanating from the locale's half-a-billion-year-old Lower Cambrian deposits, and a magnificent collection of the site's *Olenellus* and *Mesonacis* trilobites (some up to 15 centimeters in length) has been housed in the Smithsonian ever since. By the late twentieth century, however, it was believed that the Parker Quarry had either been played out by overzealous digging in the previous century or had simply been lost to the whims of modern development. But in early 2012 a small team of determined trilobite enthusiasts began studying old reports regarding the quarry's possible location. They subsequently undertook a detailed on-site investigation to discover whether the layers still existed and if they could once again be explored. It required more than two years of on and off visits to rural Vermont before an outcrop of what was believed to be the original Parker Quarry was discovered—located squarely in an active cow pasture.

MICHASPIS LIBRATA (JERGOROVO AND SAVITZKY, 1968)
Middle Cambrian; Mayan Stage; Anabar Plateau; Sakha Republic, Siberia, Russia; 4.8 cm

KINZERS FORMATION, PENNSYLVANIA

The Kinzers Formation in Lancaster County, Pennsylvania, has long been renowned for its distinctive Lower Cambrian trilobite fauna, much of which hails from a series of closely aligned quarries that have been explored and collected since the late nineteenth century. Of specific interest are the beautifully preserved examples of *Olenellus getzi* that emanate from a half-a-billion-year-old sedimentary exposure near the local hamlet of Rohrestown. These specimens, which can produce impressive examples exceeding 20 centimeters in length, are often preserved in a bright yellow or orange limonite, and they occasionally display soft-tissue preservation—mostly antennae—as well as wonderfully detailed dorsal morphology. Another Kinzers species near and dear to the hearts of trilobite hobbyists is *Wanneria walcottana*, named in honor of the seemingly ubiquitous Charles Walcott. Although complete *Wanneria* specimens are now rarely found (the area's Brubaker and Getz quarries have been closed since the 1970s, with the former now covered by a parking lot), in the late twentieth century these large, elegant fossil arthropods were unearthed in significant numbers, subsequently becoming staples of museum and private collections throughout the northeastern United States.

JINCE FORMATION, CZECH REPUBLIC

Perhaps no trilobite location on planet Earth possesses more Paleozoic gravitas than the Czech Republic's legendary Jince Formation—a series of predominantly Middle Cambrian outcrops that cover roughly 300 square kilometers in the heart of what was once known as Bohemia. Much of the formation's notoriety stems from the groundbreaking efforts performed in that area nearly two centuries ago by the French naturalist Joachim Barrande. During nearly 40 years of work, Barrande discovered and described scores of trilobite species, including *Paradoxides gracilis, Conocoryphe cirina*, and *Ellipsocephalus hoffi*. Many of these discoveries served as the centerpieces for his historic, multivolume work *Système Silurien du Centre de la Bohême*, an effort that would come to influence many, including Charles Darwin. The region's trilobites are most often found in a heavily compacted, chocolate-colored mudstone that, when properly cracked open, can occasionally reveal beautifully preserved positive/negative splits. These distinctive outcrops have been slowly exposed over multimillennia as the meandering Berounka River cuts through the tree-lined

Litavka Valley near the small town of Jince. In fact, that community of 2,200 is so proud of its world-spanning paleontological renown that it has adopted an illustrated and somewhat stylized rendering of a trilobite as a central component of its official city flag.

MANUELS RIVER, NEWFOUNDLAND, CANADA

In 1874, trilobites were first discovered near Conception Bay, Newfoundland, by a survey team working under the auspices of the Geological Survey of Canada. However, by the time renowned trilobite enthusiast Riccardo Levi-Setti stumbled upon this then long-abandoned Middle Cambrian outcrop in the mid-1970s, the rocky expanse along the Manuels River had degenerated into a makeshift garbage dump. After removing the rusting hulks of refrigerators and washing machines, Levi-Setti was able to uncover a 510-million-year-old mudstone layer filled with magnificent examples of large *Paradoxides davidis*, along with a smattering of additional trilobite species, such as *Anopolenus henrici*. Not only were these specimens aesthetically pleasing (as showcased in Levi-Setti's subsequent trilobite books that frequently highlighted this self-collected Manuels River material), but they also proved to be of singular scientific importance. Indeed, when compared to nearly identical *Paradoxides davidis* specimens found within the similarly aged Cambrian strata of Wales (now some 3,200 kilometers away), these trilobites provided key support to the concept of plate tectonics, which explains the movement of continents over time. Trilobite collecting is now banned at this legendary location, but a small museum featuring some of Levi-Setti's most notable discoveries has been opened near the Manuels River site.

(OPPOSITE PAGE) *REDLICHIA TAKOOENSIS* LU, 1950

Lower Cambrian, Middle Series 2; Botoman equivalent stage;
Emu Bay Shale; Kangaroo Island, Australia; Larger trilobite 6.5 cm

(TOP, RIGHT) *TRICREPICEPHALUS TEXANUS*
(SHUMARD, 1861)

Middle Cambrian; House Range; Weeks Formation; Millard
County, Utah, United States; 5.6 cm

(BOTTOM) *KUANYANGIA PUSTULOSA* LU, 1941

Lower Cambrian; Chengjiang Biota, Maotianshan Shales;
Heilinpu Formation; Haikou village, Chengjiang County, Yunnan
Province, China; 7.1 cm

(LEFT) *WANNERIA WALCOTTANA (WANNER, 1901)*

Lower Cambrian; Kinzers Formation; Lancaster County, Pennsylvania, United States; 6.8 cm

(BOTTOM) *ORYGMASPIS (PARABOLINOIDES) MORRISI* CHATTERTON AND GIBB, 2016

Upper Cambrian, Furongian; McKay Group, Taenicephalus Zone; Cranbrook, British Columbia, Canada; 4.4 cm

2

10 WORLD-CLASS FOSSIL SHOWS

These days fossil shows are omnipresent—held in highway-hugging motels in the American Southwest, in tiny towns on picturesque French mountainsides, and in the heart of bustling metropolises in Australia and Asia. Equal parts carnival sideshow, fossil swap meet, and scientific convention, these natural-history-inspired events manage annually to draw tens of thousands to their clarion call, often making them the most prominent, exciting, and lucrative tourist attraction their surrounding communities experience all year. And whether you're young or old, rich or poor, a London slick or a rural hick, you're welcome to attend . . . and to spend. To the uninitiated, these geologically inspired shows can appear to be one small step removed from organized chaos. Beer flats full of rare tourmaline crystals from Brazil rest gingerly upon broken-down tables, and interested observers jostle one another within an adjacent room's narrow confines to be the first to grab a particularly precious trilobite such as *Bristolia bristolenis* from California or *Breviredlichia granulosa* from China. For most showgoers, such a bizarre *and* bazaar atmosphere is all just part of the fun—although many attendees would grudgingly admit that there is a fine line separating love from hate when it comes to dealing with the out-of-control aura that can, at times, surround the fossil show experience.

Here's a look at 10 of the planet's top fossil shows.

TUCSON, ARIZONA

Each year nearly 100,000 visitors attend this city's famed fossil and mineral show, marking it as the largest event of its kind in the world. Of course, the fact that the Tucson extravaganza traditionally plays out over a three-week period in early February, whereas most other such shows—which usually take place in summer or fall—are generally limited to three- to seven-day runs, helps this Arizona exposition stand out as the planet's most impressive natural history gathering in both attendance and profitability. Quite simply, if you want to see the latest and greatest in the natural history field, you *must* go to Tucson at least once in your life.

MUNICH, GERMANY

Well-documented stories indicate that as much as 20 million euros can change hands during a typical Munich Mineral, Gem, and Fossil Show, which makes it the largest such event by far to take place on the European continent. Many of the more notable transactions—which range from the sale of complete mammal skeletons from Messel, Germany, to partial remains of the legendary *Archaeopteryx* from nearby Solnhofen—happen late at night, behind closed doors, in small crowded rooms lined with carefully displayed rows of exotic trilobites, opalized dinosaur bones, world-class minerals, and glittering ammonites. Held on the periphery of town every October, most of the trilobites found at the Munich show emanate from Morocco, whose dealers find the relatively short

(OPPOSITE PAGE) ***BRISTOLIA BRISTOLENSIS* (RESSER, 1928)**

Lower Cambrian; Latham Shale Formation; California, United States; 9.3 cm

haul (at least when compared to Tucson or Tokyo) very much to their liking.

DENVER, COLORADO

In many ways, fossil and mineral shows like the one held over a week's time every September in Denver, Colorado, are the natural history scene's equivalent of the United Nations, or at least the Tower of Babel. A stroll down any dimly lit motel aisle or tent-lined outdoor causeway is liable to bring you face-to-face with merchants from China, Germany, Morocco, Australia, France, Russia, Japan, Canada, Bolivia, and the United States, each of whom seems determined to reflect their nation's ancestral ability to either hard sell or soft sell the variety of glittering geological goods they have at their disposal. Some of these merchants expect, even want, to haggle over price (in any number of languages), though others appear to display outright disdain toward even considering such an unsavory practice.

SPRINGFIELD, ILLINOIS

Presented annually by the Mid-American Paleontological Society (MAPS), this relatively small, fossils-only show is one of the most eagerly anticipated events on the U.S. natural history calendar. Although international visitors have long wondered why those who run this three-day, paleo-packed production insist on holding it in the ostensible middle of nowhere (actually, in the town of Springfield, Illinois, with a population of 115,000), longtime MAPS participants thrive on knowing that the show's rural roots and somewhat rustic attitude keep away some of the "city slickers" who drive up both costs and prices at larger shows held in more centralized cities around the globe.

SAINTE-MARIE-AUX-MINES, FRANCE

Taking place each summer in the beautiful Alsace region of eastern France, the Sainte-Marie fossil and mineral show is perhaps the most charming such event in the world. Despite the verdant natural scenery and the free-flowing wine (drawn from myriad local vineyards), just successfully navigating the intricate maze of delicate mineral crystals and fragile spiny trilobites that line many of the exhibition's crowded indoor galleries and overflowing tent pavilions is a perpetual challenge. You may first be gingerly stepping over an assemblage of Jurassic-age Czech reptiles left haphazardly on the floor and then circumventing a chest-high display of 100-million-year-old German mosasaur bones that one dealer has decided to lean against his car door. But these hassles are well worth the price of attending this outstanding event because traditionally the trilobites—especially those from Europe and Morocco—are plentiful.

BEIJING, CHINA

One of the newer and more impressively staged events on the natural history show schedule, initial reviews of the Beijing Mineral, Gem, and Fossil Show have been—to be polite—somewhat mixed. Many attendees found Chinese security to be not only expectedly tight but also unnecessarily intimidating and somewhat overwhelming. Indeed, most show participants coming from distant lands said that merely getting their material into and then back out of this often mystifying nation proved to be a major headache. Still, those who managed to overlook these significant inconveniences report that the Chinese show promoters did everything possible to turn their gathering into one of the largest, most all-consuming events of its kind, an

exposition that seems certain to grow in both size and prestige in the coming years.

SPRINGFIELD, MASSACHUSETTS

The most impressive fossil show in the northeastern United States is held annually in Springfield, Massachusetts, a midsized city located equidistant—approximately 200 kilometers—between the thriving metropolises of New York and Boston. Unlike similar exhibitions presented in Shanghai, Munich, and Denver, where such natural history events are generally absorbed by either the size of their host city or the less-centralized location of the show setting, the fossil and mineral spectacular in Springfield has a major impact on both the community's attitude and its economy. Many east coast dealers who have little inclination to undertake the long hike to Tucson or Denver (let alone to an international destination) rely on the brisk business provided by the fossil-rich Springfield event.

TOKYO, JAPAN

Held in the lobby and side rooms of a major downtown hotel, and making up for its limited size by the outstanding quality of its featured dealers, Tokyo's annual International Minerals and Fossils Show is one of the most prestigious natural history events in the world. Indeed, a mere invitation to display one's fossiliferous wares at this show immediately serves to move any merchant's industry standing up a major notch or two. In Tokyo, visitors and collectors can easily get up close and personal with the objects of their Deep Time desires. You can pick up and examine just about any available piece, but be careful—if you break it, you take it, and some of the displayed trilobites (particularly those from Russia) are tagged at $10,000 (1,400,000 yen) or more! Precious few indigenous trilobites can be found on this beautiful volcanic island, so most of

BREVIREDLICHIA GRANULOSA ZHANG AND LIN IN YIN AND LEE, 1978

Lower Cambrian, Series 2, Stage 4; Wulongqing Formation (Guanshan Biota); Yunnan Province, China; 9.3 cm

the arthropod material hails from North America and Morocco, although in recent years more and more Chinese dealers have begun making their Paleozoic presence felt.

EDISON, NEW JERSEY

For years, if not decades, fossil lovers living in and around New York City complained about the lack of a world-class fossil show in their immediate vicinity. Oh, sure, such finicky folks could readily hop on a plane and travel to Tucson, Denver, or even Tokyo. They could even undertake the three-hour drive up to Springfield. But what about something in their own proverbial backyard? Well, such an event (which began in 2014) lasted for two years in the shadow of the Big Apple before being shifted to Edison, New Jersey—named after the prolific inventor Thomas Edison—a small town located some 90 kilometers southwest of downtown Manhattan. There a smattering of select east-coast-centric fossil dealers have helped transform the annual New Jersey Fossil and Mineral Show into a significant stop on the U.S. show schedule. One may have to look hard to find them, but trilobite treasures can be uncovered, many coming from the relatively nearby and ever-abundant Paleozoic outcrops of upstate New York, Pennsylvania, and southern Ontario.

STUTTGART, GERMANY

This small, well-run event held on the outskirts of the midsized German city of Stuttgart (population 648,000) is rapidly emerging as one of Europe's "must see" natural history gatherings. Some have compared this self-proclaimed Fossil Market to the MAPS show in the United States due to its fossils-only guidelines (no bothersome jewelry, mineral, or tchotchke dealers) and select show participants. Many European merchants who are hesitant to undertake the monster-sized trek—and expense—associated with attending major North American or Asian shows are content to save some of their best material for display at this Fossilien Borse. These merchants know that many of the European continent's top trilobite enthusiasts will be in attendance, and they hope the pockets of these collectors will be brimming with plenty of ready to spend cash.

(LEFT) *LICHAS MAROCANUS* DESTOMBES, 1968
Upper Ordovician; Ktaoua Formation; Tazarine, Morocco; 12.4 cm

(BOTTOM) *GABRICERAURUS SP.*
Middle Ordovician; Vaureal Formation;
Eastern Quebec, Canada; 13.3 cm

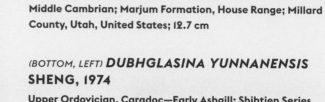

(TOP, LEFT) OLENOIDES INFLATUS (WALCOTT, 1909)

Middle Cambrian; Marjum Formation, House Range; Millard County, Utah, United States; 12.7 cm

(BOTTOM, LEFT) DUBHGLASINA YUNNANENSIS SHENG, 1974

Upper Ordovician, Caradoc—Early Ashgill; Shihtien Series, Pupiao Formation; Shidian, Baoshan, Yunnan Province, China; 4.1 cm

(BOTTOM, RIGHT) FAILLEANA SP.

Upper Ordovician; Poolville Formation; Criner Hills, Oklahoma, United States; 8 cm

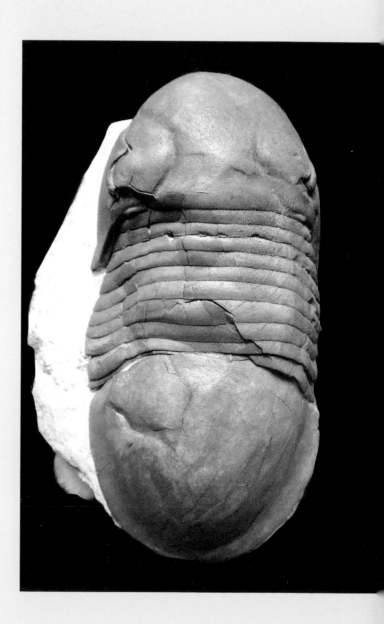

10 OUTSTANDING ORDOVICIAN TRILOBITE LOCALITIES

485 to 444 Million Years Ago

Soon after the dawn of the Ordovician Period, approximately 485 million years ago, trilobites reached what many academics now consider their evolutionary peak, producing thousands of new species, including *Isotelus latus, Parisoceraurus rectangulatus*, and *Amphilichas ottawaensis*. Some 40 million years into their incredibly successful swim through the Paleozoic Era, these highly adaptable arthropods clearly remained one of the world's dominant forms of marine life. But that dominance was beginning to be threatened by the emergence of an ever-increasing army of predators—including giant cephalopods and primitive jawed fish, many of whom viewed the neighboring trilobite population as little more than a convenient source for a quick and easy protein snack. With all of the continents then located in Earth's southern hemisphere, most Ordovician trilobite life was congregated around the equatorial regions along the northernmost coasts of these landmasses. Much of what would eventually emerge as North America and Europe was then covered by warm, shallow seas, a fact that allowed trilobites to eventually become fossilized in locations now hundreds of kilometers from the nearest body of salt water. By the end of that period, some marine creatures had become bold enough to venture permanently onto land, beginning to inhabit the fringes of the continental mass known as Gondwana. Things seemed to be evolving at a sprightly pace both in the sea and on land, but disaster loomed again. The second major extinction event at the end of the Ordovician—this one possibly

***PARISOCERAURUS CF. RECTANGULUS
ZHOU, 1977***

Upper Ordovician; Pupiao Formation; Baoshan region; Yunnan
Province, China; 7.7 cm

were noteworthy for their thick, caramel-colored calcite shells as well as for their outstanding three-dimensional preservation. Some specimens, including *Paraceraurus exsul* and *Niobe schmidti*, featured exotic shapes and sizes. Others, such as *Hoplolichas furcifer*, presented carapaces covered with a virtually impenetrable assemblage of pointed spines. Equally as impressive, many of the more than 100 trilobite species discovered and described from this fossiliferous horizon had previously been unseen by the scientific community, at least in their current levels of completeness. By the mid-1990s, these magnificent arthropods were being widely marketed at trade shows and exhibited in museums, with one in particular—*Asaphus kowalewskii*—drawing special attention because it featured eyes sitting atop stalks often 5 centimeters long.

generated by rapid global cooling—eradicated virtually all terrestrial organisms and severely depleted life in Earth's oceans as well.

The following sections describe 10 outstanding Ordovician trilobite localities.

ASERY HORIZON, RUSSIA

Following the fall of communism in the late-1980s, a veritable trilobite tidal wave began to emerge from the Volkhov River region near St. Petersburg and inundate the global market. The preponderance of this Russian arthropod material was being drawn from a series of geologically associated layers collectively known as the Asery Horizon, which had formed 450 million years ago when this area of Eastern Europe was still covered by a warm inland sea. These Ordovician-age trilobites

WALCOTT/RUST QUARRY, NEW YORK

The Ordovician-age Walcott/Rust Quarry of central New York State owes its name, as well as much of its fame, to two of American paleontology's most renowned figures: Charles Walcott, who decades later would play a vital role in the discovery of British Columbia's Burgess Shale, and William Rust, who spent the final 40 years of his life uncovering this locale's expansive Paleozoic fauna. At last count, 18 species of trilobites have been found in the finely grained, 455-million-year-old limestone deposits that dominate this picturesque tree-lined site. These discoveries include massive plates of *Ceraurus pleurexanthemus* and *Flexicalymene senaria*, along with football-sized examples of *Isotelus gigas*, each featuring a magnificent black calcite preservation. Thought to have been lost to science at the tail end of the nineteenth century, a late-twentieth-century rediscovery sparked renewed interest in both this legendary quarry

***NIOBE SCHMIDTI* (BALASHOVA, 1976)**

Middle Ordovician; Kunda level—Obuchov Formation; Voybokalo, St. Petersburg region, Russia; 7.3 cm

and its equally celebrated trilobite material. In fact, during the initial decades of the twenty-first century, after a major commercial dig was begun at the site, prime Walcott/Rust specimens have continually been among the star paleontological attractions at various fossils shows held from Tucson to Tokyo.

FILLMORE FORMATION, UTAH

Nowhere on the face of the North American continent are trilobite fossils more prevalent than in the majestic state of Utah. Ranging from the 510-million-year-old Middle Cambrian Wheeler Shale, through the 440-million-year-old Ordovician-age Fillmore Formation, this western outpost brims with some of the most acclaimed and studied Paleozoic outcrops in the fossil record. The region's Cambrian trilobites include such familiar species as *Olenoides nevadensis, Modocia typicalis*, and *Hemirhodon amplipyge*, as well as the world's most common trilobite, the pervasive *Elrathia kingii*. Ordovician species include *Trigonocercella acuta, Presbynelius ibexensis, Bathyurellus teretus*, and *Isoteloides flexus*. Yet for all the paleontological work that has been conducted in Utah over the last 150 years—with more than 500 trilobite species being revealed—hidden amid the countless canyons and valleys that cut through the Drum, House, and Wellsville mountain ranges are previously unknown fossil-bearing locations, along with a corresponding array of new trilobites that are still being discovered on a regular basis.

VALONGO FORMATION, PORTUGAL

Portugal's 450-million-year-old Valongo Formation preserves key elements of a temperate marine ecosystem that has produced some of the largest trilobite specimens ever unearthed, with many of these supersized arthropods exceeding 40 centimeters in length. Various stone slabs found within the famed Louzeiras de Canelas Quarry feature hauntingly preserved images of rare species such as *Placoparia tournemini* and *Hungioides bohemicus*. Other sedimentary surfaces present nearly a dozen large *Ogyginus forteyi* trilobites lying virtually one atop the other, with their fossilized carapaces filling the surrounding matrix with unmistakable evidence of their long-ago lives. Since the late nineteenth century, both commercial workers and residents near the town of Arouca have extracted the area's Ordovician-age slate blocks primarily for use as roofing and paving tiles. In recent years, however, the quarrying operations have hit a surprising fossil-rich layer that has yielded trilobites of prodigious sizes, and often in prodigious numbers. Appearing as ghostly alabaster images that contrast dramatically against the jet-black shale, more than a score of trilobite species have been described from this notable outcrop of the Valongo Formation, and many of these have become world renowned for their beauty, size, and scientific importance.

GIRVAN, SCOTLAND

Lady Burn has long been recognized as one of Europe's most revered, studied, and diverse Ordovician fossil localities. The area that encompasses this site adjacent to the small Scottish town of Girvan lies along that country's rugged western coast, about 100 kilometers south of Glasgow. It is a rustic environment filled with rolling green hills, rock-strewn outcrops, and well-maintained farms—some still controlled by the British royal family. For centuries, area residents have walked through this charming yet challenging terrain while keeping a sharp eye peeled for the unique

BATHYURELLUS TERETUS YOUNG, 1973
Middle Ordovician; Fillmore Formation, Pogonip Group; Millard County, Utah, United States; 5.1 cm

"fossil pockets" that characterize the legendary locale's 437-million-year-old deposits. These finely grained outcrops of hard, calcareous sandstone can yield well-preserved examples of starfish, crinoids, and bryozoans, in addition to more than two dozen distinct trilobite species. Lady Burn fossils are often coated with a sturdy brown or orange mineralized patina that serves to instantly distinguish the site's array of unusual trilobite genera. These indigenous examples include _Toxochasmops bissetti_, _Pseudosphaerexochus octolobatus_, and the giant cheirurid _Hadromeros keisleyensis_, which on rare occasions has been found as a complete specimen up to 25 centimeters in length.

ONTARIO, CANADA

The various 460-million-year-old Ordovician Period formations that emerge throughout the southern region of Ontario, Canada—which include the Lindsay, Cobourg, Bobcaygeon, and Verulam—rank high in any compendium featuring the world's leading trilobite sites. More than a century of exploration and excavation has been conducted on the area's fossil-saturated outcrops both by professionals hailing from Toronto's nearby Royal Ontario Museum and by dedicated legions of local amateur enthusiasts. During that time these varied sedimentary layers have provided the collecting community with some of the most impressive, fascinating, and scientifically significant trilobites ever brought to light. Such notable species as *Ceraurinus marginatus, Isotelus gigas, Ectenaspis homalonotoides,* and *Hemiarges paulianus* hail from the diverse yet geologically associated formations of the region. Specimens are usually three-dimensionally preserved and have a thick, dark-brown calcite shell. Unfortunately, in recent years many of the area's quarries have been closed due to safety concerns, making Ontario's already available trilobite material even more precious to those who truly appreciate the special morphological qualities exhibited by these amazing arthropods.

FEZOUATA FORMATION, MOROCCO

Deep in the heart of the arid wilderness that characterizes much of southern Morocco, and located amid a series of rugged sedimentary outcrops that represent what was once the floor of a shallow Ordovician sea, lies the 478-million-year-old Fezouata Formation. Since its scientific discovery in the late 1990s, this incredibly rich Paleozoic ecosystem has yielded more than 2,000 invertebrate specimens representing 50 distinct taxa—many being of nonbiomineralized organisms that strongly resemble those found within the 30-million-year-older Burgess Shale of western Canada. The faunal similarities seen within these two formations—separated by both age and distance—distinguish this deposit as a unique and vital link between "holdover" genera that managed to exist in both Middle Cambrian and Lower Ordovician marine environments. Fezouata's trilobites share little in common with known Burgess species. They abound within the formation's chalky limestone layers, with such distinctively detailed species as *Megistaspis hammondi, Foulonia peregrina,* and *Parvilichas marochii* emerging as the focused subjects of both scientific study and

collector interest. On occasion, these intricately fossilized trilobites even display evidence of legs, antennae, and internal organ preservation.

MOUNT ORAB, OHIO

Rarely does a trilobite receive recognition as an official state fossil. As might be expected, dinosaurs enjoy this designation in numerous regions across the United States, as do the occasional woolly mammoth and saber-toothed cat. But in Ohio, the large *Isotelus maximus* specimens that for more than a century have been pulled from the 447-million-year-old Ordovician deposits that sit atop Mount Orab have earned that exact distinction. However, these impressive Arnheim Formation examples, which on occasion reach 45 centimeters in length, are not the only renowned species that emerge from these rocks; *Flexicalymene retrorsa* rank as the formation's most prolific trilobites, and the beautiful and rare *Allolichas halli* represents one of the true trilo-trophies of the American fossil landscape. The formation also features an impressive assortment of crinoids, cephalopods, and brachiopods, all of which lurk amid the relatively soft gray limestone layers. It is the giant ovate *Isotelus,* however, with their smooth, chocolate-brown calcite carapaces, that continue to draw trilobite enthusiasts to Mount Orab year after year in hopes of finding the "ultimate" complete specimen.

BEECHER'S TRILOBITE BED, NEW YORK

For more than 120 years, *Triarthrus eatoni* trilobites of unique beauty and academic importance have been unearthed from within the thinly striated Upper Ordovician layers of the Lorraine Shale in Oneida County, New York. What makes these small trilobites (usually 3 centimeters or less in length) particularly remarkable is not only their dramatically pyritized preservation but the fact that many of these 450-million-year-old fossils present exceptional soft-tissue detail. Ever since Charles Emerson Beecher of Yale University began work at this site during the last decade of the nineteenth century, the formation's trilobites have been renowned for displaying the finest examples of ventral morphology ever found. Indeed, when carefully prepared, many specimens provide evidence of legs, antennae, gills, and even eggs. It was long believed that Beecher had exhausted the location during his initial three-year dig, and upon his death in 1904 it was assumed that the outcrop, commonly referred to as Beecher's Trilobite Bed, had been lost forever. However, in 1984 the amateur paleontologists Thomas Whiteley and Dan Cooper—subsequently and notably followed by the researcher/preparator Markus Martin—rediscovered the location, and ongoing excavations

PSEUDOSPHAEREXOCHUS OCTOLOBATUS (McCOY, 1849)
Upper Ordovician, Ashgill Series; South Threave Formation, Farden Member, "Starfish Bed"; Girvan, Ayrshire, Scotland, United Kingdom; 5.7 cm

have since revealed hundreds of additional complete *Triarthrus* specimens, along with some rare unidentified species of lichids and odontopleurids.

PUPIAO FORMATION, CHINA

During the first two decades of the twenty-first century, a veritable deluge of exciting, previously unseen, and generally unknown trilobite material began emerging from the Paleozoic deposits of China. These often exotic specimens—some exhibiting well-defined antennae and appendages—were soon filling internet websites and highlighting displays at rock and mineral shows held around the globe. Such well-received public appearances made it abundantly clear to trilobite enthusiasts from Beijing to Boston that the entire Asian continent was dotted with an impressive diversity of fossiliferous formations—dating from the Lower Cambrian to the Upper Devonian—many of which had hitherto managed to avoid detection by scientifically inclined eyes. Yunnan Province's now legendary Lower Cambrian Chengjiang Biota understandably grabbed a solid share of academic acclaim, but other locations—such as Yunnan's Upper Ordovician Pupiao Formation—quickly proved that their diverse fauna also warranted the attention of both academics and collectors. Often preserved as three-dimensional internal molds that appear as yellow, mineralized "stains" on a brick-red mudstone, midsized Pupiao Formation trilobites such as *Parisoceraurus rectangularis, Metopolichas sp.,* and *Dubhglasina yunnanensis* (most examples between 3 and 6 centimeters in length) have been found amid this area's thickly banded layers of rough-hewn rock.

(OPPOSITE PAGE) ***CERAURINUS MARGINATUS (BARTON, 1913)***

Upper Ordovician; Cobourg Formation; Colborne, Ontario, Canada; larger specimen 6.2 cm

FOULONIA PEREGRINA (DEAN, 1966)

Lower Ordovician, Floian Stage; Upper Fezouata Formation; Zagora area, Morocco; largest specimen 6 cm

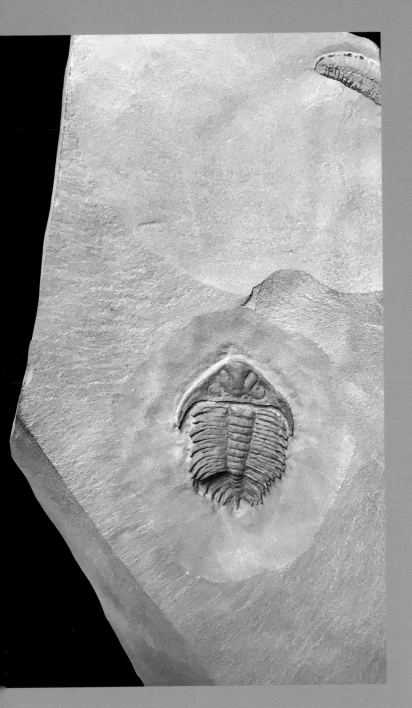

(LEFT) LICHID N. SP.

Upper Ordovician, Maysvillian Regional Stage; Whetstone Gulf Formation, Lorraine Group; Martin Quarry; Beecher's Trilobite Bed; Oneida County, New York, United States; 4.2 cm

That's a partial *Triarthrus* in upper-right corner.

(BOTTOM) ODONTOPLEURID N. SP.

Upper Ordovician, Maysvillian Regional Stage; Whetstone Gulf Formation, Lorraine Group; Martin Quarry; Beecher's Trilobite Bed; Oneida County, New York, United States; 2 cm

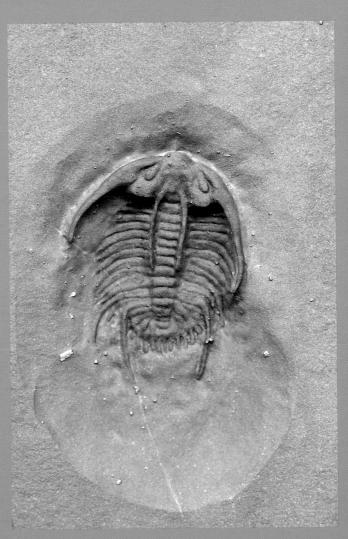

***PRIONOCHEILUS PULCHER* (BARRANDE, 1846)**

Lower Ordovician; Ktaoua Formation; El Kaid Errami Region, Morocco; 7.2 cm

10 ESSENTIAL FIGURES IN TRILOBITE HISTORY

Either as living creatures or fossilized forms, trilobites have been part of our world for more than 520 million years, but the scientific study of these ancient arthropods has only been conducted in earnest throughout the last few centuries. During that time, a select few adventurers and academics have risen above all others in their quest to unearth the remains of these Paleozoic creatures and to unravel the role trilobites played in Earth's history. Some of our Ice Age ancestors in Europe apparently revered trilobites, as did a variety of Native American tribes, especially those located in the southwestern desert of the United States. Members of the Ute tribe routinely carried small, 510-million-year-old *Elrathia kingii* specimens in their medicine pouches to provide protection from enemies and to ward off evil spirits. Despite these early manifestations of trilo-centric interest, however, the true study of these primal invertebrates didn't begin until the last years of the seventeenth century. It was then, in England, that the Reverend Edward Lhuyd made the first direct mention of a trilobite in scientific literature, describing something he called a "flat fish," which later would be identified as the trilobite *Ogygiocarella debuchii*.

Half a century later, in 1750, the Englishman Charles Lyttleton conducted the initial research on a specific trilobite species, submitting a paper to the Royal Society of London on the famed Dudley Locust, now known as the Silurian trilobite *Calymene blumenbachii*. Nearly another century would go by before the renowned Scottish geologist Roderick Murchison laid the groundwork for future trilobite studies with his historic volume *The Silurian System*. That manuscript,

CYRTOMETOPUS CF. CLAVIFRONS (DALMAN, 1827)

Middle Ordovician; Lower Elnes Formation, Helskjer Member; Furnes Village, Ringsaker Municipality; Innlandet County, Norway; 5.1 cm

released in 1839, intricately described the fossil fauna found throughout Britain, and its ensuing notoriety turned Murchison into a sensation in Europe's most erudite circles. Murchison's celebrated efforts led directly to the oft-lauded trilobite-related research conducted by the likes of the French naturalist Joachim Barrande and the American adventurer Charles Walcott. Their accomplishments, along with subsequent endeavors by many others, including contemporary paleontologists such as Niles Eldredge and Richard Fortey, have added important chapters to the evolving story of the world's favorite fossilized arthropod.

Here are the 10 essential figures in trilobite history.

CHARLES WALCOTT (1850–1927)

From his initial exploration of upstate New York's Ordovician strata—in what would eventually become known as the Walcott/Rust Quarry—to his legendary work at British Columbia's Burgess Shale, few figures in the history of paleontology come close to matching the lifetime achievements of Charles Walcott. The renowned expert of all things Middle Cambrian during his lifetime, his accomplishments remain unrivaled among those who have sought knowledge of the primeval past.

JOACHIM BARRANDE (1799–1883)

This French naturalist spent decades during the mid-nineteenth century collecting and studying trilobites within the borders of what was then known as Bohemia, now the Czech Republic. His groundbreaking, multivolume masterpiece, *Système Silurien du Centre de la Bohême*, not only unveiled previously unimagined information about trilobites but also presented wonderfully detailed, hand-drawn images of the specimens he discovered. Barrande's seminal work supposedly even influenced the thinking of Charles Darwin.

SIR RODERICK MURCHISON (1792–1871)

In the middle years of the nineteenth century, Murchison emerged as a true "rock star" geologist throughout the British Isles. He helped pioneer the study of trilobites in his groundbreaking book *The Silurian System*, the success of which turned him into a national icon who appeared in front of thousands of rapt followers whenever he gave one of his fossil-themed lectures. In 1863, he was knighted for his sundry paleontological achievements.

JAMES HALL (1811–1898)

Due to the far-reaching impact of his 1847 volume *The Invertebrate Fossils of New York*, perhaps no one played a more vital role than James Hall in establishing the science of paleontology in the United States. During his time as the state geologist of New York, his work allowed him to explore, as well as collect, a wide variety of key trilobite-bearing horizons, including the Rochester Shale. His

influence on other scientists (at different times, both Charles Walcott and Charles Beecher served as apprentices) helped shape the contemporary approach toward the study of invertebrate fossils.

LOUIS AGASSIZ (1807–1873)

This Swiss-born geologist and biologist is perhaps best known for founding Harvard's Museum of Comparative Zoology. Agassiz subsequently filled that institution with one of the major trilobite collections in the world—including both the best Barrande-collected material to make it out of the

CHANCIA EBDOME (WALCOTT, 1924)

Middle Cambrian; Langston Formation; Spence Shale; Utah, United States; 3.5 cm

First identified by the legendary Charles Walcott.

Czech Republic and an outstanding assortment of Walcott/Rust fauna. In fact, his early support of Charles Walcott, who was then in his early 20s, spurred the famed explorer to make many of his unprecedented paleontological discoveries.

CHARLES DARWIN (1809–1882)

Although Darwin may never have directly studied or, for all we know, even *held* a trilobite, his renown as the father of modern evolutionary thought still instantly qualifies him as an essential member of our top 10 cavalcade. Indeed, it is believed that the earlier writings of various scientists whose primary focus was trilobites (including both Murchison and Barrande) directly influenced Darwin's subsequent concepts and philosophies.

HARRY WHITTINGTON (1916–2010)

Due to his mid-twentieth century efforts at the Burgess Shale in British Columbia, many within the scientific field consider Whittington to be one of the most important forces behind the widespread recognition of the Cambrian Explosion. Working first in his native Britain, and later in the United States as the curator of invertebrate paleontology at Harvard's famed Museum of Comparative Zoology, his actions helped identify the incredible faunal diversity that existed during the Cambrian.

STEPHEN JAY GOULD (1941–2002)

Prior to his passing in 2002, Gould was the renowned professor of zoology at Harvard University who helped revolutionize scientific thought through his pioneering work (with Niles Eldredge) on the theory of punctuated equilibria. That concept describes how evolutionary change—rather than being a slow, steady process—often occurs in sudden bursts. A prodigious writer with more than 300 mainstream and academic articles to his credit, his book on the Burgess Shale, *Wonderful Life*, is a must-read for anyone interested in trilobites and the Cambrian Explosion.

RICHARD FORTEY (1946–)

This British scientist stands as perhaps the modern "face" of trilobite research. Fortey's various books on the subject (including Y2K's *Trilobite: Eyewitness to Evolution*) have served to introduce an ever-wider audience to the myriad wonders of these amazing arthropods. In addition, his pursuits at the Natural History Museum in London—where he worked for more than three decades—have expanded our knowledge of how these primal organisms lived and interacted in the undersea world that surrounded them.

NILES ELDREDGE (1943–)

The curator of invertebrate paleontology at New York's legendary American Museum of Natural History from 1969 until his retirement in 2010, it was in 1972 that Eldredge (along with Stephen Jay Gould) proposed one of the most celebrated scientific theories of the late twentieth century: punctuated equilibria. Much of Eldredge's work in that regard was based on his studies of the compound eye patterns of North American phacopid trilobite species, some of which have subsequently been given the genus name *Eldredgeops* in his honor.

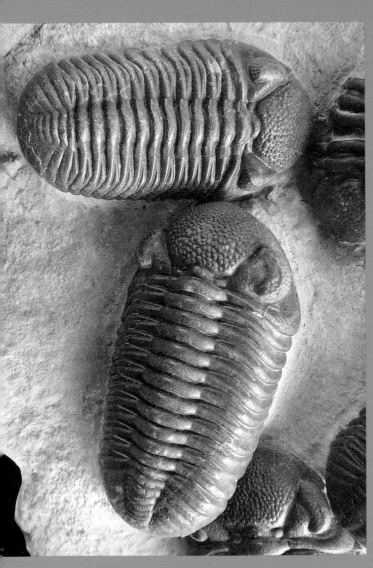

(LEFT) *ELDREDGEOPS MILLERI* (STEWART, 1927)

Middle Devonian, Givetian; Silica Shale Formation; Milan, Michigan, United States; largest trilobite 7.3 cm

Named after the paleontologist Niles Eldredge.

(BOTTOM) *LLOYDOLITHUS LLOYDI* (MURCHISON, 1839)

Lower Ordovician; Llandeilo Series; Meadowtown Beds; Betton, Shropshire, England; 6.3 cm with spines

The famed Scottish geologist Rodrick Murchison first identified this classic species.

OGYGINUS FORTEYI (RABANO, 1989)

Lower Ordovician; Valongo Formation; Arouca, Portugal; 37.8 cm

This large species is named in honor of the British paleontologist Richard Fortey.

5

10 SIGNIFICANT SILURIAN TRILOBITE LOCALITIES

444 to 419 Million Years Ago

The Silurian Period began 444 million years ago and witnessed a rejuvenation of planetary life following the mass extinction that had occurred at the end of the Ordovician. With a duration of *only* 25 million years, this was the shortest and perhaps least extraordinary of the various Paleozoic periods. Unlike the Cambrian, which holds special scientific cachet for bearing witness to the initial flowering of multicellular life, or even the Devonian, during which myriad familiar-looking creatures began to rear their often bizarre heads, the Silurian was a period marked most notably by global climatic stabilization and a significant rise, then fall, in worldwide sea levels. Trilobites continued their inexorable decline, which began after their generic peak in the Lower Ordovician, but these incredible invertebrates still produced thousands of strikingly diverse new species. These included *Arctinurus boltoni*, *Aanasobella asper*, and *Metopolichas breviceps*, each of which displayed the ever-increasing degree of morphological complexity these arthropods were forced to assume merely to survive in their chosen marine ecosystems. This frequently spectacular array of trilobites continued to thrive in Earth's oceans—which then included the Iapetus and Panthalassic—and were joined in that undersea realm by an ever-growing assembly of fish, as well as various species of eurypterids, some of which were among the apex predators of the Silurian. After a series of unsuccessful early attempts, life returned to the land, and plants began to dominate the global terrain. At this moment in time, four distinct continents existed on Earth, and these along with their surrounding waters provided ample

***CHEIRURUS INFENSUS* CAMPBELL, 1967**

Middle Silurian; Henryhouse Formation, Hunton group; Pontotoc County, Oklahoma, United States; 3.2 cm

new ecological environments in which diverse life-forms, including trilobites, could continue to emerge and evolve.

Here's a look at 10 significant Silurian trilobite repositories.

HENRYHOUSE FORMATION, OKLAHOMA

Often overshadowed in the minds of collectors and academics by the impressive assortment of Devonian trilobites that spring from the shales of the neighboring Haragan and Bois d'Arc formations, the Silurian-age Henryhouse Formation is also a prolific source of prime Paleozoic material. Located amid the rolling Arbuckle Mountains in rustic Pontotoc County, Oklahoma, this site's 430-million-year-old trilobite fauna has long been treasured for its rarity, beauty, and scientific significance. Much like their world-renowned Devonian cousins, these Silurian Sooner State bugs are traditionally three-dimensionally preserved in a thick, caramel-hued calcite that contrasts strikingly against the surrounding buff-colored matrix. Indigenous species include *Dalmanites rutellum* (almost always discovered in an arched or enrolled state), the pustulose *Fragiscutum glebalis*, the rare cheirurid, *Cheirurus infensus*, and the formation's most common trilobite, *Calymene clavicula*, examples of which can be found up to 6 centimeters in length. Although the abundance and diversity of Henryhouse material has never proven equal to that of the legendary Haragan Formation, the Henryhouse has produced an impressive 11 trilobite species encompassing 10 distinct genera.

WREN'S NEST, DUDLEY, ENGLAND

The Silurian exposures of the English Midlands have been yielding amazing trilobite specimens for centuries, with those coming from the Much Wenlock limestone formations near the town of Dudley rating among the most beautiful and renowned in the entire fossil domain. One particular species, *Calymene blumenbachii*, has become synonymous with the region, and the city of Dudley's town crest proudly boasts an illustrated version of this fossilized local "locust." More than 80 trilobite species—including *Cybantyx anaglyptos*, *Dalmanites myops*, *Acaste inflata*, and *Encrinurus tuberculatus*—have been discovered and described from the area's most famous collecting locale, Wren's Nest, where the 425-million-year-old strata has produced thousands of magnificently preserved complete specimens. In the mid-nineteenth century, enthusiasm for the Wren's Nest site was overwhelming, and when the noted Scottish geologist Sir Roderick Murchison (author of the groundbreaking 1839 volume *The Silurian System*) lectured on Dudley's fauna, his talks often attracted stunningly large throngs of enthusiastic listeners. One such gathering, held at Wren's Nest itself, drew an estimated crowd of 15,000 fossil-frenzied folks and generated national headlines. Now designated as a Site of Special Scientific Interest (where digging is prohibited), Dudley remains among the most studied and storied of European fossil locales.

ROCHESTER SHALE, MIDDLEPORT, NEW YORK

The fossils of the Silurian-age Rochester Shale were originally noted during construction of the Erie Canal, a project that carried on continuously from 1814 until 1825. At that time, discovery of the area's 420-million-year-old trilobites—including *Trimerus delphinocephalus*, *Arctinurus boltoni*, *Dalmanites limulurus*, and *Dicranopeltis nereus*—generated as much confusion and consternation as answers among member of the scientific community. Quite simply, no one had previously seen a treasure trove of Paleozoic material quite like this, and few (if any) knew what to make of these unexpected fossil riches. In fact, some of the earliest scientific descriptions of many invertebrate species were based solely on discoveries made within this historic locale in western New York State. The site has long been famous for its outstanding preservation and abundance of invertebrate life; indeed, the Rochester Shale features more than 200 described species, including corals, bryozoans, brachiopods, and bivalves—as well as more than a dozen recognized trilobite species.

ANTICOSTI ISLAND, QUEBEC

One of the most prolific Silurian-age trilobite-bearing zones in North America can be found beneath the rugged yet picturesque landscape of remote Anticosti Island, Quebec. The island is surrounded by the surging waters of the Gulf of

BALIZOMA CF. VARIOLARIS (BRONGNIART, 1822)
Middle Silurian, Wenlockian Stage, Homerian Substage; Much Wenlock Limestone Formation; Wren's Nest Hill, Dudley, West Midlands, England; 4.2 cm

St. Lawrence, so it takes a hearty and determined soul to reach this densely forested, decidedly out-of-the-way locale. But such a venture is well worth the effort for those with a true Paleozoic passion. The 1,800-meter-thick sedimentary strata of Anticosti is brimming with fossils—in some spots they're quite literally spilling out of the cliffsides that surround the island. Many of the trilobite specimens discovered—including *Diacalymene schucherti*, *Failleana magnifica*, and *Arctinurus anticostiensis*—provide ample evidence that some 430 million years ago this marine ecosystem was home to megafauna, including trilobites of unusual size (some over 20 centimeters in length) and with spectacular preservation. After a major 2004 scientific revision, 52 trilobite species are now recognized as coming from Anticosti's rich Silurian layers, and these comprise an impressive 30 genera. Although material from formations such as the Jupiter and Gun River may not be as well-known to collectors as trilobites from more renowned North American repositories, as more academic expeditions head to this craggy outpost (commercial exploration is now banned), Anticosti Island's status as a trilobite haven seems destined to steadily increase.

GRAFTON, ILLINOIS

In certain Paleozoic locations around the planet, trilobites emerge from their eons-old sedimentary encasements as little more than dull, dolomite-infused internal molds—specimens totally devoid of their attractive calcite carapaces. One such site occurs near the small midwestern town of Grafton, Illinois, where for decades three-dimensionally preserved Silurian trilobites of various sizes, shapes, and species have been culled from the ivory-toned Niagaran limestone outcrops that comprise the surrounding Joliet Formation. Since dedicated amateur collecting first

began at this locale in the 1950s, thousands of trilobites have been recovered, often by those wading through 10-foot-high piles of weathered scree. By far the most prominent of these dolomitic bugs are the 425-million-year-old examples of *Calymene celebra*; due to their availability and inexpensive price of acquisition, they have become an expected component of virtually every major fossil collection. But far more unusual trilo-types, including *Cerauromerus hydei*, *Dalmanites platycaudatus*, *Encrinurus egani*, and *Ommokris vigilans*—the latter two species featuring eyes sitting atop prominent stalks—have been found in complete form at this renowned site, marking it as one of North America's most prominent Silurian exposures.

WALDRON SHALE, INDIANA

Abundant sedimentary outcrops of Silurian age rank among the relative rarities of the fossil record. For trilobite enthusiasts, such widely recognized formations as England's Wren's Nest and New York's Rochester Shale stand among the premier Silurian fossil repositories on the planet. Another notable but perhaps lesser known U.S. locality lies within the Waldron Shale of Indiana, where beautifully preserved 425-million-year-old examples of such unusual trilobite species as *Metopolichas breviceps*, *Glyptambon verrucosus*, and *Trimerus delphinocephalus* (some of which exceed 12 centimeters in length) have been found in outcrops that dot the state's southern half. Perhaps closest in faunal content to the Rochester Shale, there are still marked differences in the material uncovered in the slightly younger strata of the Waldron, and the fossils here—which also include an array of brachiopods, crinoids, and gastropods—appear to be less prevalent than in the New York locale. Because of their rarity, the attractive appearance of their mottled calcite carapaces, and their lifelike preservation, Waldron trilobites rank as particularly prized acquisitions by hobbyists around the globe.

GOTLAND, SWEDEN

On a large island situated nearly 90 kilometers off Sweden's southeastern coast, trilobites have been excavated and studied from Gotland's 420-million-year-old sedimentary deposits since 1851. Almost a millennium earlier, this 2,100-square-kilometer refuge served as an important Viking trading settlement, remnants of which can still be seen hidden amid the area's rugged, boulder-strewn shores. In key spots throughout Gotland, thick limestone beds have perfectly preserved a rich Silurian fauna featuring an abundant array of trilobite species. Usually fossilized in a fine, toffee-colored calcite, these wonderfully inflated, although generally diminutive, specimens include such trilobites as *Proetus granulatus*, *Kettneraspis angelini*, *Sphaerexochus latifrons*, and *Calymene neotuberculata*, none of which exceed 4 centimeters in length. Often compared in both age and preservation to the renowned Much Wenlock outcrops of the English Midlands, Gotland specimens are perhaps even more prized due to their relative scarcity. With many of the fossil-bearing Gotland layers now completely submerged under the waters of the surrounding Baltic Sea, recently unearthed examples from this locale are few and far between.

XIUSHAN FORMATION, CHINA

Until the late 1990s, it was a relatively common occurrence to encounter many hard-at-work Chinese fossil dealers when strolling through the massive rock and mineral extravaganzas that take place annually in Tucson, Denver, Munich, and Tokyo. In such show settings, the displays presented by these stalwart merchants would often be crammed with a hefty supply of Asian fossiliferous material, ranging from Triassic *Keicheosaurus* plates and Cretaceous *Hadrosaur* eggs, to a slew of choice Cambrian, Ordovician, and Silurian trilobites. Alas, for a variety of reasons, both sensible and nonsensical, these days those same Chinese dealers are rarely spotted outside of their home country. Despite their twenty-first-century fossil show scarcity, at least a smattering of the fruits generated by their initial labors remain in paleontological play. Lower Silurian trilobites from the Xiushan Formation of Guizhou Province (first introduced internationally by these merchants a quarter of a century ago) have become omnipresent on the internet, with species such as the common *Coronocephalus gaoluoensis* and the very rare *Senticucullus elegans*—the latter featuring a ring of prominent spikes encircling its cephalon—drawing the ardent attentions of trilobite collectors from Shanghai to Springfield.

COTTON FORMATION, AUSTRALIA

Australian trilobites have long held a special cachet among collectors. Some of that almost mystical allure stems from a basic geographical fact—the considerable distance that separates this famed island continent from the living rooms of most Paleozoic hobbyists. Throughout the last half-century, the preponderance of collector focus has been cast upon the startling assortment of Lower Cambrian material hailing from South Australia's famed Emu Bay Shale Formation. However, trilobite enthusiasts have slowly begun to note the presence of an equally compelling arthropod assemblage emerging from the Lower Silurian deposits of New South Wales. There, within the thinly striated limestone bands that mark the 420-million-year-old Cotton Formation, such diminutive (2- to 4-centimeter-long) species as *Raphiophorus sandfordi*, *Odontopleura (Sinespinaspis) markhami*,

***NUCLEURUS DEOMENOS* TRIPP, 1962**

Lower Silurian; Jupiter Formation; Anticosti Island, Eastern Quebec, Canada; 4.6 cm

and *Aulacopleura pogsoni* have been brought to light, and these have slowly begun making their way onto the world market, often as intriguing positive/negative splits. Many of the species drawn from the Cotton Formation outcrops have been known since the middle of the twentieth century, but it wasn't until 2001 that scientific recognition was awarded to these trilobites, virtually all of which have been unearthed as well-preserved internal molds. These discoveries have also provided science with a greater understanding of the primal forces that have long worked to shape the face of the planet, especially since a surprisingly similar array of Silurian species has been uncovered in such now disparate locales as the Czech Republic, northwestern Canada, and South China.

MOTOL FORMATION, CZECH REPUBLIC

When one of the most heralded, studied, and respected books in paleontological parlance is called *Système Silurien du Centre de la Bohême*, it shouldn't be the least bit surprising to learn that some of the Czech Republic's most notable trilobite-bearing deposits are indeed Silurian. Of course, it can't be totally ignored that the preponderance of trilobites in the first and perhaps most notable volume of Joachim Barrande's lauded late-nineteenth-century work are actually Cambrian and Devonian in age. However, throughout this picturesque land once known as Bohemia are more than enough Silurian outcrops to make any trilobite enthusiast sit up and take serious notice. One of the more prominent of these fossil-laden sites is the Motol Formation, located near the small town of Lodenice (population 1,991), where thick layers of dark-brown, fauna-bearing mudstone emerge from the surrounding hillsides at intermittent intervals. Some of these sedimentary sheets are covered in fossiliferous residue ranging from graptolites to (on much rarer occasions) complete trilobites. Over the course of centuries, such highly prized species as *Dicranopeltis scabra propinqua, Decoroproetus decoratus*, and *Diacalymene diademata* have been unearthed from these revered 430-million-year-old outcrops, with many of the recovered examples subsequently forming the core Silurian collections housed in some of Eastern Europe's most respected natural history museums.

(TOP) *CERAUROMERUS HYDEI* (WELLER, 1907)

Lower Silurian; Niagaran Series, Joliet Formation; Grafton, Illinois, United States; 3.8 cm

(RIGHT) *ODONTOPLEURA (SINESPINASPIS) MARKHAMI* EDGECOMBE AND SHERWIN, 2001

Lower Silurian, Late Llandovery Series, Telychian; Cotton Formation; Cotton Hill Quarry, Forbes, New South Wales, Australia; 1.3 cm

***DICRANOPELTIS SCABRA PROPINQUA* (BARRANDE, 1846)**

Middle Silurian; Motol Formation; Lodenice, Czech Republic; 4.4 cm

6

10 KEY ELEMENTS OF TRILOBITE MORPHOLOGY

From a morphological perspective, all trilobites appeared as variations on a surprisingly similar theme. Despite existing for more than a quarter of a billion years and producing tens of thousands of scientifically recognized species (with that number steadily increasing on an annual basis), for the most part trilobites shared a strikingly uniform body plan. There was good reason for these shell-deep similarities: even at a very early stage in the history of life on our planet, the trilobite model had already proven it possessed a certain degree of evolutionary perfection. Unquestionably, not all trilobites looked the same. Some species had eyes, some didn't. Some trilobites reached impressive dimensions, others did not. Many presented long, flowing genal spines, though others displayed none at all. Indeed, during their lengthy trek through time, trilobites existed in an almost dizzying assortment of shapes and sizes. Perhaps no other animal class in Earth's history has exhibited the diversity of primal design shown by these amazing arthropods. But at their heart (and yes, trilobites apparently did possess a basic but effective cardiovascular system) they were all remarkably alike. When you look at the fossilized remains of a trilobite, whether it is a 6-centimeter-long Cambrian-age *Olenellus fowleri* or a 10-centimeter Ordovician *Isotelus iowensis*, please consider that lurking under those imposing (and occasionally spine-encrusted) calcite carapaces were animals that represented one of our planet's first and most successful experiments with complex biological life. From their initial moments on Earth some 521 million years ago, few creatures have ever been as evolutionarily "perfect" in their various anatomical

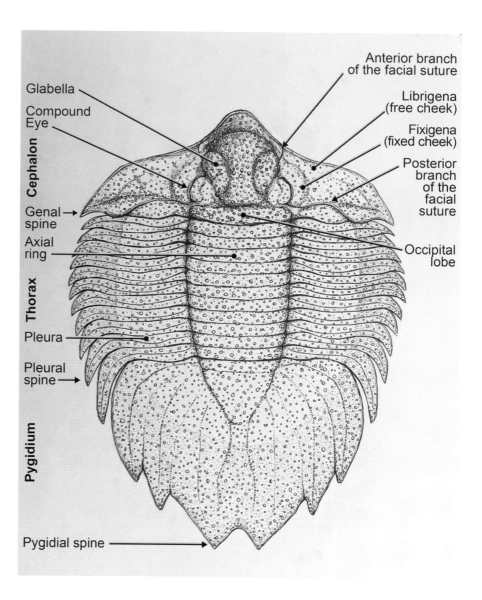

Glabella
Compound Eye
Cephalon
Genal spine
Axial ring
Thorax
Pleura
Pleural spine
Pygidium
Pygidial spine

Anterior branch of the facial suture
Librigena (free cheek)
Fixigena (fixed cheek)
Posterior branch of the facial suture
Occipital lobe

DRAWING BY BEN COOPER

aspects as that fascinating organism known as the trilobite.

The sections that follow describe 10 key elements of trilobite morphology.

CEPHALON

Trilobites were named not, as is generally surmised, for their three primary body segments—cephalon (head), thorax (body), and pygidium (tail)—but rather for the three lobes that longitudinally divided their dorsal exoskeleton. Yet it is

their easily recognizable cephalon, which features the creatures' eyes, suture lines, and glabella (and which are often found as tantalizingly disarticulated trilo-parts upon many properly aged sedimentary surfaces), that is perhaps most easily noted by those interested in the fossils of these often bizarre marine inhabitants.

THORAX

Although trilobites have no direct living relatives, they do share many of the definitive characteristics

ISOTELUS IOWENSIS (OWEN, 1852)

Middle Ordovician; Maquekota Formation; Bowling Green, Missouri, United States; Larger trilobite 11.3 cm

of the arthropod phylum—including a multisegmented and flexible thorax—with contemporary creatures ranging from brine shrimp to wood lice. Yet despite these superficially similar characteristics, key details of the trilobites' primeval anatomy, including aspects of their thoracic articulation, mark their class's members as a totally separate line on Earth's family tree.

PYGIDIUM

The diverse shapes of their pygidia can be as varied as the trilobites themselves. Scores of species present tails featuring ornate pleural furrows, though others are as smooth as a proverbial baby's bottom. Still others exhibit strange telson-like extensions that occasionally come close to exceeding the entire length of their bodies. Some trilobite pygidia had evolved to look like the back end of a hydrodynamic missile, whereas others made their host trilobite resemble nothing more than a primordial puff pastry.

OPISTHOTHORAX

A small but significant number of early trilobite species presented a strange, multisegmented opisthothorax, an elongated anatomical feature that

EXOSKELETON

The trilobites' hard outer shell is unquestionably the most instantly identifiable aspect of their dorsal anatomy. After all, it is the fossilized form of that distinctive, calcite-coated carapace that collectors so intently seek to pry from select sedimentary stratum around the globe. Whether preserved in some sort of mudstone outcrop or in various slates, shales, or sandstones, those characteristic trilobite exoskeletons are the most tangible evidence yet uncovered that reflect the long-ago lives led by these primeval arthropods.

EYES

It has been said that the Lower Cambrian appearance of arthropod eyes ranks among the most significant evolutionary developments in the history of life on our planet. From the magnificently detailed schizochroal eyes that adorn Devonian phacopid species such as *Eldredgeops milleri* to the amazingly complex holochroal lens systems that characterize the optics of scutellids like *Spiniscutellum umbeliferum*, the eyes of these incredible invertebrates represent a truly unique morphological feature, one that no creature, either before or since, has managed to equal in its calcite-infused complexity.

SPINES

As predators in those primal Paleozoic seas continued to become both more impressive and more aggressive, the trilobite response to such menacing marine threats was a biologically logical one—to grow protective spines of significant length, thickness, and sharpness. This notable degree of trilobite spinosity dates all the way back to the Lower Cambrian, but such morphological extravagance apparently reached its peak 120 million years later

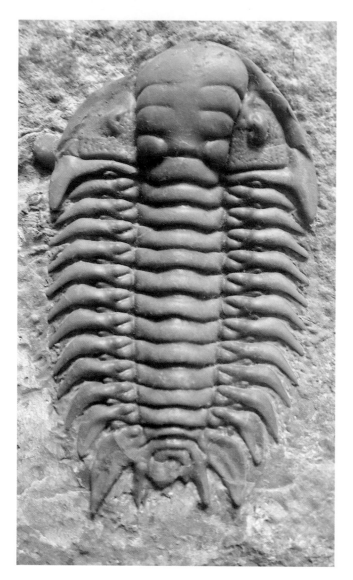

"XYLABION" SP.
Upper Ordovician; Bobcaygeon Formation; Simcoe County, Ontario, Canada; 7 cm

extended from the creature's thorax and provides possible evidence of the trilobite line's even more primitive, wormlike predecessors. Lower Cambrian trilobites such as Nevada's *Olenellus fowleri* are renowned for their extended opisthothorax, and Australia's diminutive species *Balcoracania dailyi* holds the trilobite record with nearly 100 thoracic segments, including a pronounced opisthothorax.

in the Middle Devonian, a time when some phacopid species evolved into creatures most resembling spine-covered undersea war machines.

FACIAL SUTURES

Facial sutures—located on either side of the trilobite's cephalon—allowed these arthropods to more easily shed their hard, calcite exoskeleton during times of molting. The earliest trilobite species apparently did not benefit from this evolutionarily advantageous anatomical feature, but by the Middle Cambrian—some 12 million years into their swim through Deep Time—virtually all of these creatures were capable of rooting their pygidia upon the seafloor and with a bit of thoracic twist to pop the top of their shell off along the suture lines. They were then able to wriggle out of the resultant opening and patiently wait for a new, larger carapace to grow.

HYPOSTOME

The trilobite hypostome, or mouth plate—which in many modern arthropods is indicative of a parasitic lifestyle—was present in all trilobite genera, although the shape of this morphological feature varied rather radically from species to species. In some, it appeared along the trilobite's ventral side as a rounded, dual-sided "prong," and in others it was more of a forklike projection that could grow to be more than half the length of the trilobite's body. Academics now believe that the hypostome may have served sundry purposes for the host trilobite: helping to scrape food from surrounding rock surfaces, allowing the animal to attach itself to marine structures (such as coral reefs) during storms, and perhaps even serving as a form of mating clasper.

ANTENNAE

Dating to early studies done by Charles Walcott in the late nineteenth century—conducted on both Cambrian and Ordovician species—it was firmly established that trilobites possessed a plethora of nonbiomineralized soft-parts, including legs, gills, and antennae. In more recent years, trilobite material emerging from both the Lower Cambrian Chengjiang Biota of China and the Lower Ordovician Fezouata Formation of Morocco has often presented remarkably well-preserved specimens featuring prominent and unmistakable antennae.

***TRIARTHRUS EATONI* (HALL, 1838)**

Upper Ordovician; Lorraine Shale, Martin Quarry; Beecher's Trilobite Bed, Oneida County; New York, United States; 0.9 cm

Photo courtesy of Markus Martin

(LEFT) NIOBELLA SP.

Lower Ordovician, Tremadoc; Hunan Province, China; 7.3 cm

(BOTTOM) OLENELLUS FOWLERI (PALMER, 1998)

Lower Cambrian; Pioche Shale Formation; Nevada, United States; 6.8 cm

URIPES SP. NOV.

Upper Ordovician, Ashgillian; Quarrel Hill Formation, Glenmard Member; Glenmard Quarry, south side of Quarrel Hill; Girvan, Ayrshire, Scotland; 3.7 cm

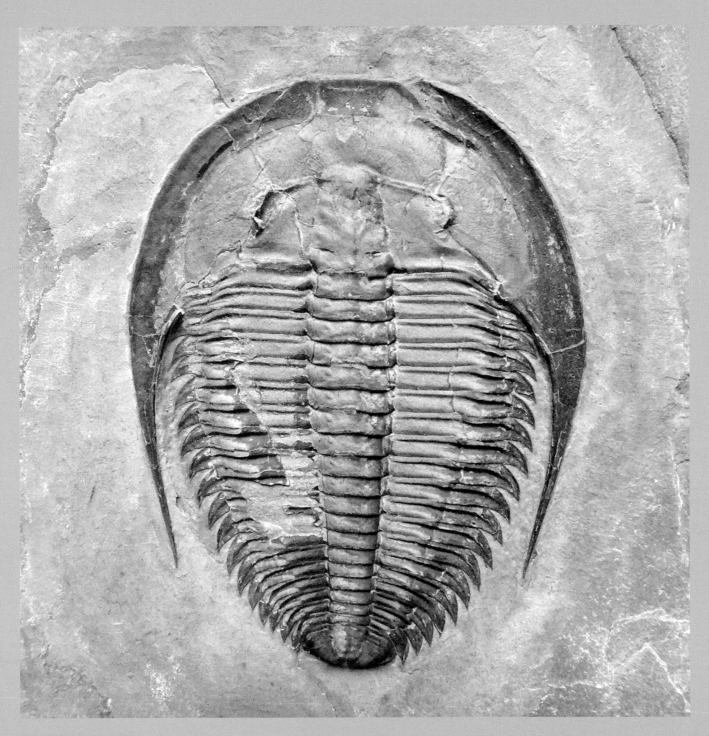

AMECEPHALUS SP.

Middle Cambrian, Miaolingian Series, Wuliuan Stage; Langston Formation, Spence Shale Member; Box Elder and Cache counties; Utah, United States; 6.8 cm

PROCERATOCEPHALA SP.

Upper Ordovician; Lower Ktaoua Formation; Jebel Tijarfaïouine; El Kaid Errami, Morocco; 4.1 cm

7

10 DRAMATIC DEVONIAN TRILOBITE LOCALITIES

419 to 359 Million Years Ago

By 419 million years ago the Devonian was underway, a period when huge armor-plated fish—some up to 9 meters in length—first roamed the seas. These imposing creatures posed a further evolutionary hurdle to the remaining trilobites, which then included *Huntoniatonia huntonensis*, *Dicranurus monstrosus*, and *Belenopyge balliviani*. The Devonian's cast of giant predators, which also featured terrorizing sea scorpions and jet-propelled cephalopods, even threatened trilobite species that had emerged specifically to combat these lurking marine menaces, such as the spike-covered *Drotops armatus*. Although in sharp decline, trilobites continued to produce morphologically more complex species, many featuring an intriguing—and often intimidating—assortment of flowing spines, pointed barbs, and imposing rostrums, each of which seemingly served a vital function in the animal's ability to swim, crawl, feed, defend itself, or reproduce. Trilobite eyes—as shown on species such as *Eldredgeops milleri*—also reached their apex during the Devonian, frequently appearing atop the trilobite's rounded cephalon as huge, compound ocular outlets covered by hundreds of perfectly formed geometrically interconnecting lenses. During this period, plants continued to monopolize the land, helping to develop a more hospitable oxygen-filled environment that enabled terrestrial life to slowly begin to flourish. But once again Earth proved a fickle host, and the planet's next great extinction event eradicated 70 percent of life around the globe—a disaster that emphatically and dramatically signaled the end of the Devonian.

**CHOTECOPS FERDINANDI
(KAYSER, 1880)**

Lower Devonian; Hunsruck Slate;
Bundenbach, Rhine Valley, Germany;
6.2 cm

Here is a look at 10 of the planet's most significant Devonian outcrops.

HUNSRUCK, GERMANY

Since the fourteenth century, thin, finely grained sheets of charcoal-gray slate (used primarily as roofing tiles) have been pulled from Devonian-age quarries located throughout the mountainous German region of Hunsruck. Despite a series of world wars and economic downturns, quarrying at the site, commonly referred to in fossil-centric circles as Bundenbach due to that town's nearby proximity, has continued virtually unabated. Although certainly not the primary reason for this ongoing mining activity, the appearance of a magnificent and scientifically important fossil fauna has repeatedly drawn international attention to this Rhine Valley repository. Perhaps best known for its diverse assemblage of 400-million-year-old starfish and crinoids, even the enticing evidence of jawless fish has made occasional appearances within the area's centimeter-thick Hunsruck Slate sheets. The trilobites found here include *Chotecops ferdinandi, Rhenops lethaeae,* and *Parahomalonotus planus,* with the *Chotecops* being the most prolific by far. These specimens are also notable for their occasional soft-tissue preservation; and when subjected to modern X-ray technology, they have yielded amazing anatomical detail, including evidence of antennae, gills, and walking legs.

ALNIF, MOROCCO

Over the last three-plus decades, Alnif, Morocco, has grown into a somewhat unlikely epicenter for cutting-edge trilobite discoveries, as well as for the subsequent entry of these specimens onto the active commercial market. This small, desert-hugging town (population 3,000) has provided the world's collecting community with a veritable bonanza of bizarre genera ranging from Lower Cambrian fallotaspids—among the first recognized trilobites, over 520 million years old—through an incredible array of unusual Middle Devonian species, many, such as *Dicranurus monstrosus*, *Quadrops flexuosa*, and *Metacanthina barrandei*, featuring outlandish spines or huge, multifaceted eyes. In recent years, scores of previously unknown species have been recovered from the freshly opened quarries that dot the surrounding fossil-rich landscape. Some of these exotic examples require the delicate hands of skilled preparation artisans to free them from their eons-old rock encasements—work still often performed without the benefit of a reliable electrical source or the proper array of matrix-extracting tools. A dearth of scientific research has been performed on the preponderance of material found within Morocco's expanding inventory of Devonian exposures. But a growing number of paleontologists have recently begun turning their attention toward the daunting yet fascinating task of analyzing the myriad morphological complexities presented by this overwhelming army of time-spanning arthropods.

HARAGAN FORMATION, OKLAHOMA

Throughout the last 30-odd years, the fossil-filled Devonian sediments of Oklahoma's Haragan Formation have produced an amazing array of magnificent, three-dimensional trilobite specimens. A rich caramel color, and often featuring strange spinose ornamentation (most notably on the species *Dicranurus hamatus elegantus* and *Ceratonurus sp.*), these 417-million-year-old trilobites—found atop a legendary outcrop known in paleontological circles as Black Cat Mountain—have long ranked as favorites for trilobite enthusiasts around the globe. With their lifelike preservation and intricate surface textures, these ancient relics are beautiful enough to serve as the star attractions in just about any major museum exhibit or private trilobite display. Indeed, some 20 species are known from the Haragan and its sister formation, the Bois d'Arc—including *Huntoniatonia huntonensis*, *Kettneraspis williamsi*, *Reedops deckeri*, and the rare *Acanthopyge consanguinea*—making this hill-strewn, semiarid region of south-central Oklahoma one of the most prolific trilobite-producing localities in North America. Ironically, many of the same genera found at this Oklahoma site are mirrored in the Devonian rocks of Morocco, knowledge that has lent significant academic support to the theory of plate tectonics.

LA PAZ, BOLIVIA

Since the middle years of the nineteenth century, a unique assemblage of trilobite material has been recognized from a series of closely aligned Devonian formations, including the Sicasica, Belen, and El Carmen, all of which ring the outskirts of La Paz, Bolivia. These outcrops are located at a dizzying height of 3,600 meters atop that nation's famed

***HUNTONIATONIA HUNTONENSIS* (ULRICH AND DELO, 1940)**

Lower Devonian; Haragan Formation; Black Cat Mountain Quarry; Clarita, Coal County, Oklahoma, United States; 11.4 cm

Altiplano, marking this as one of the most inaccessible and intriguing trilobite-bearing localities on the planet. The region's 370-million-year-old fossils are almost always found in hard limestone concretions, which occur when mineral-rich elements housed within the sedimentary strata attach to and then condense around a static object, such as a trilobite carapace. These specimens have long been a favorite of collectors around the globe as well as being the subject of intense scientific study. Dozens of trilobite species—including *Dipleura boliviensis*, *Maurotarion legrandi*, *Viaphacops orourensis*, *Eldredgia venustus*, and *Belenopyge balliviani*—have been identified from these prolific zones, with many emerging from their eons-old rock encasements as beautifully preserved positive/negative splits. A significant number of these 3- to 8-centimeter-long trilobites are indigenous only to the area's mid-Paleozoic exposures. Their special status is apparently due to the isolated geographic position once held by what is now the soaring Altiplano. Back in the Devonian that entire region served as the floor of a marine shelf that had become separated from other thriving coastal communities.

EIFEL REGION, GERMANY

Middle Devonian trilobites were initially found in the Eifel region of Germany's Rhine Valley in 1825, when the geologist Heinrich Georg Bronn discovered a small, semienrolled specimen he named *Calymene schlotheimi*, a species that has subsequently been reclassified as *Geesops schlotheimi*. Since that time, hundreds of complete and partial examples of that phacopid have been unearthed in this prolific outcrop of the 395-million-year-old Ahrdorf Formation. In fact, the fossil-oriented excavations at this site near the town of Gees became so intense during the second half of the twentieth century—including, on occasion, the use of powerful rock-moving machinery—that in 1984 German officials decided to intervene and ban all further commercial digging. By then, however, a full assortment of beautifully preserved, three-dimensional trilobites representing dozens of species—including such rarities as *Asteropyge punctata*, *Ceratarges armata*, and *Scutellum geesense*—had already been recovered. These 2- to 5-centimeter, brown to black trilobites continue to be considered the gems of many European museum displays—and are the much-coveted centerpieces of private collections around the globe.

SILICA SHALE, OHIO

Few trilobite sites in the world have continually stirred the souls of collectors more than the Silica Shale of Sylvania, Ohio. For more than a century, this area's Middle Devonian quarries have been noted for their fossil-infused fauna: in addition to countless pyritized brachiopods, intricate crinoid stems, and detailed coral calices, this area features three trilobite species—two closely related phacopids (*Eldredgeops milleri* and *Eldredgeops crassituburculata*) and one rare proetid (*Basidechenella lucasensis*). But it is not the rather limited number of trilobite species found in this locale's coarse, light-gray mudstone matrix that excites collectors. Rather, it is the spectacular preservation of the specimens uncovered there—wonderful three-dimensional examples covered in a thick, charcoal-gray calcite that seems to capture every nuance of trilobite dorsal morphology. Unquestionably among the most alluring and notable of these anatomical aspects are *Eldredgeops*'s huge compound eyes, magnificently preserved with every optic lens and geometric detail intact. With a little imagination, those eyes almost appear to be staring back at you, providing a dramatic link to life more than 400 million years in the past.

CORTEZ MOUNTAINS, NEVADA

In recent years, a high elevation sedimentary exposure housed in the Lower Devonian Wenban Limestone of Nevada's Cortez Mountains has yielded a modest supply of spectacular trilobite material. These specimens include complete, exceptionally large examples of rare phacopids, lichids, odontopleurids, and dalmanitids—some yet to be scientifically described. Since the latest exploration of the Wenban's 415-million-year-old layers began in 2009, the recovered trilobite material has been nothing less than remarkable in both size and quality, if not necessarily quantity. Approximately two dozen complete examples of the spinose, 12-centimeter-long *Viaphacops claviger* have been found, along with three specimens of the 18-centimeter *Synphoria nevadensis*, which features a bizarre "inverted" tail. But it is the two distinctly different species of unidentified lichids—one with a bulbous glabella and a series of lengthy spines ringing its pygidium and the other with pronounced eyes and a huge, half-moon-shaped tail—that have attracted the preponderance of interest from both collectors and academics. Some of these lichid specimens are up to 40 centimeters in length, placing them among the largest Devonian trilobites ever unearthed in North America.

PENN-DIXIE, NEW YORK

A select number of fossil-filled locations that border Lake Erie in western New York State have long been noted for their exceptional Middle Devonian fauna. Indeed, reports regarding the area's 385-million-year-old crinoids, brachiopods, and bivalves date back nearly two centuries. Despite the attention those abundant Paleozoic remnants have continually drawn from the local populace, however, it is the unique accumulation of trilobites that can be found in the region's rich Windom Shale and Moscow Formation limestone deposits that has continually attracted a lion's share of acclaim. Both Eighteen Mile Creek (named for its distance from the Niagara River) and the nearby Penn-Dixie quarry are of singular paleontological interest due to their beautifully preserved examples of trilobites such as *Eldredgeops rana*, *Dipleura dekayi*, *Bellacartwrightia calliteles*, and *Greenops barberi*. The 2- to 4-centimeter-long *Eldredgeops* are of special note because they can occasionally be unearthed in mass-mortality plates featuring dozens of overlapping specimens, a phenomenon

BELENOPYGE BALLIVIANI PEK AND VANEK, 1991
Middle Devonian; Belen Formation; La Paz, Bolivia; 5.9 cm

some scientists believe may represent mating or molting assemblages. In addition, these locations have proven to be extremely popular with amateur fossil collectors who have been known to fill both Penn-Dixie and Eighteen Mile Creek—usually in organized "club" outings—during weekend digs held regularly from May until October.

JORF FORMATION, MOROCCO

Over the last half-century no nation on the planet has produced more trilobite fossils than Morocco. Yet despite the thousands of complete and often spectacular Cambrian, Ordovician, and Devonian specimens that have emerged from this fossil-packed North African hotbed, interested observers have speculated that only a small fraction of the region's available trilobite resources have been fully explored. To further support such conjecture, most of the Moroccan quarries that were first mined in the 1980s and 1990s have now been virtually tapped out, but it is known that fresh Paleozoic horizons are still being discovered on an annual basis. These new excavations include an exciting Middle Devonian quarry located near the centrally situated town of Jorf. This outcrop first rose to prominence in 2016 with the emergence of perfectly preserved examples of the rare lichid genus *Acanthopyge*. Since then, the number and variety of these Jorf trilobites—including previously unknown species of *Harpes, Proetus*, and *Koneprusia*, many featuring a distinctive, glasslike preservation—have continued to expand, with most quickly finding their way into the appreciative hands of both the collecting and the academic communities.

ONONDAGA LIMESTONE, PENNSYLVANIA

Trilobites, as well as the geological formations in which they are contained, rarely—if ever—recognize state, provincial, or national borders. Thus the Devonian-age Onondaga Limestone Formation, which crisscrosses much of the Allegheny region of the eastern United States, can be found running through wide swaths of New York, Pennsylvania, and Ohio, along with significant sectors in Kentucky and Virginia. It even briefly stretches into the southern districts of Ontario, Canada. Contained within the finely grained, dolomitic sediments that comprise the Onondaga's geologic foundation are a plethora of trilobite species, including the diminutive odontopleurid *Acidaaspis calicera*, the legendary—and apparently always fragmentary—dalmanitid *Coronura aspectans*, and large examples (up to 13 centimeters) of the closely aligned *Odontocephalus aegeria* and *Odontocephalus selenurus*, nearly all of which have been collected and studied for more than a century. Often distorted by internal Earth forces, and never easy to find or extract, the trilobites of the Onondaga have long served as essential components in the collections of both leading East Coast museums and top tier hobbyists across the globe. Indeed, with many of the formation's "classic" sites now permanently closed, in recent years these trilobites have become more prized than ever by the world's collecting community.

(TOP) *METACANTHINA CF. BARRANDEI*

**Middle Devonian, Eifelian; AM Limstone; Ben Slimane, Meseta
East of Casablanca, Morocco; 5.4 cm**

(RIGHT) *DIPLEURA DEKAYI* (GREEN, 1832)

**Middle Devonian (Givetian); Skaneateles Formation, Hamilton
Group; Delphi Station Member, New York, United States; 18.1 cm**

VIAPHACOPS CLAVIGER (HAAS, 1969)

Lower Devonian; Wenban Formation; Cortez Range, Central Nevada, United States; Top specimen 9.7 cm

(RIGHT) SCABRISCUTELLUM SP.

Middle Devonian, Eifelian; Jorf, near Erfound; Morocco; 6.2 cm

(BOTTOM LEFT) VIAPHACOPS N. SP.

Middle Devonian; Belen Formation; La Paz, Bolivia; 7.3 cm

(BOTTOM RIGHT) WENNDORFIA PLANUS SCHRAUTI BASSE & FRANKE, 2006

Lower Devonian, Pragian; Merzâ Akhsaï Formation; Oued Ghris, Errachidia Province, Morocco; 13.4 cm

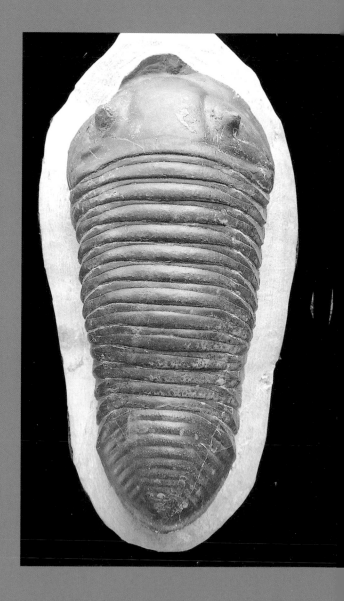

10 TRILOBITE COLLECTING TIPS

What exactly is the lure of collecting trilobites? What is it that makes grown men and women from widely divergent economic, regional, and societal backgrounds focus so intently on the Paleozoic permutations of these long-gone marine inhabitants? Of course, the concepts that trilobites were among the first dominant forms of multicellular life on Earth, that they survived for 270 million years, and that there was a seemingly endless variety of these primordial arthropods all serve as part of their ageless appeal. Perhaps it is the fact that trilobites were the earliest organisms on our planet to possess complex eyes and calcite exoskeletons, both of which are often beautifully preserved in their fossilized remains. The dizzying degree of morphological diversity that exists within the trilobites' basic three-part body plan also adds immeasurably to their inherent allure: there are the spinose lichids . . . the captivating cheirurids . . . the symmetrical asaphids . . . the streamlined harpids. And the notion that fossilized trilobite-derived debris has been found in just about every nook and cranny on Earth's craggy surface—from the distant islands of South Australia, to the burning deserts of North Africa, to rugged steppes of Eastern Siberia, to the majestic mountains of western Canada—also provides an extra element to their collectability quotient. It should also be noted that the competitive nature of these invertebrate obsessed enthusiasts—especially when it comes to being the first among peers to add the next big-eyed Moroccan phacopid or smooth-shelled Russian illaenid to their fossiliferous holdings—has helped keep the trilobite collecting

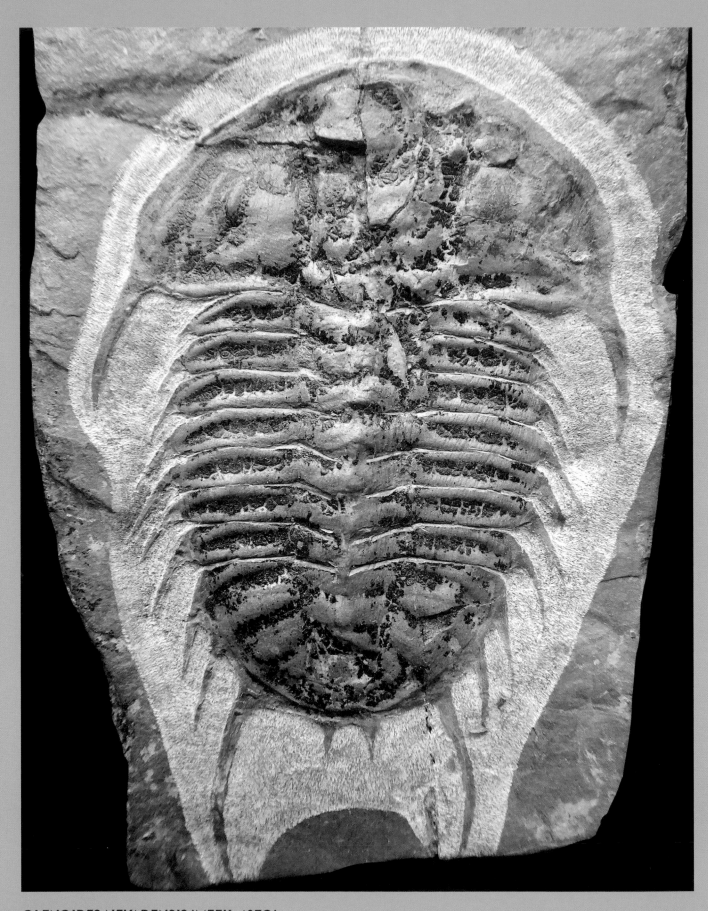

***OLENOIDES NEVADENSIS* (MEEK, 1870)**

Middle Cambrian; Wolsey Shale; Nixon Gulch; Manhattan, Montana, United States; 6.4 cm

world spinning along at a surprisingly healthy pace, from both a financial and a scholarly perspective.

With all of that in mind, here are 10 key trilobite collecting tips.

QUALITY OR QUANTITY?

One of the first—and most important—questions any trilobite enthusiast must ask concerns your prime collecting objective: Would you rather possess a score of lesser-quality bugs from a single location or a single superlative specimen from that same repository? The fact is that many hobbyists' perspectives change over time. They may start out wanting a comprehensive collection showcasing innumerable specimens from all Paleozoic time periods or all major localities, but end up realizing that assembling a select number of quality examples drawn from a specific site is more satisfying to their collecting soul.

STAY WITHIN YOUR BUDGET

There's nothing more challenging to any hobbyist than coming to the harsh realization that you have champagne tastes and beer pockets. Attending a fossil show and being surrounded by a seemingly endless (and often expensive) display of "shiny" Paleozoic baubles can frustrate even the most savvy and selective trilobite enthusiast. But it is equally essential to calculate your budgetary boundaries and to stay within those financial parameters—at least *most* of the time!

SIZE MATTERS

Most trilobite fossils are small, 7 centimeters or less from head to tail. But there's no question that larger trilobites—especially unique and well-preserved specimens over 10 centimeters in length—present an extra element of appeal for many collectors. The reason for this is relatively simple: these trilobites are easier to see and display and are more fun to show off to friends and family. There is frequently a significant size variance within examples of a specific species, for example, *Arctinurus boltoni* from upstate New York. Some of these impressive lichids may be as small as 9 centimeters, and others can range up to 16 centimeters in length—the larger examples are invariably more desired and more costly.

LOCATION

Much like the old joke about real estate, there's no question that location plays a major role in any trilobite's collectability quotient. Indeed, some enthusiasts may spend their entire collecting careers focused solely on gathering the Paleozoic output from a specific quarry, state, province, or nation. In addition, specimens of a genus such as *Dalmanites* may be relatively common from upstate New York but be as rare as a mythical trilobite tooth from Waldron, Indiana, or Dudley, England.

USE ALL AVAILABLE RESOURCES

Whether you attend fossil shows, delve deep into internet offerings, or communicate directly with others who share your Paleozoic passion, there are innumerable ways to expand your trilobite collection. In fact, more than a score of Facebook and Instagram sites are currently dedicated either partly or solely to the discussion, analysis, or sale of trilobites, which makes contacting your fellow enthusiasts an easier and more rewarding proposition for hobbyists around the globe.

SELF-COLLECTING

Of course, the most primal (and traditionally most satisfying) means of gathering trilobites for

MARJUMIA CALLAS
WALCOTT, 1916
Middle Cambrian; Marjum Formation;
Millard County, Utah, United States;
Larger trilobite 7.7 cm

a collection is by going into the field and doing the hands-on, callous causing work yourself. There you will be forced to battle against the elements—and perhaps even your digging partners—in your pursuit of Paleozoic prizes, but the results may well provide a treasured relic drawn directly from our planet's distant past.

TRADING

Some trilobite hobbyists love trading specimens with their fellow collectors, though others strongly dislike the competitive nature of such ventures. Either way, trading these incredible invertebrates—whether face-to-face, over the phone, or through emails and texts—often provides an easy, effective, and entertaining means of both reaching out to other enthusiasts around the globe and gathering hard to obtain material from Earth's distant corners.

THE INTERNET

Whether it's one of the dozens of natural-history-oriented commercial websites or the ever-present sales force known as eBay, over the last few decades the internet has proven to be an incredibly constructive and convenient means for collectors to continually expand their trilobite holdings. On occasion dark and blurry photos may mask what proves to be questionable restoration or repair, but more often than not the internet provides a relatively safe and easy source for fossil collecting fun.

FOSSIL SHOWS

As detailed elsewhere in this book, a visit to a major fossil show is something that every serious collector should experience at least once. From crates filled with a seemingly endless variety of Moroccan trilobites to glass cases brimming with the latest and greatest Paleozoic material from Russia, England, and North America, such natural history extravaganzas present a unique opportunity for enthusiasts to view (and possibly acquire) the best the fossil world has to offer. Whether that show is held in Tucson, Tokyo, or Munich, each of these annual events should be considered as nothing less than a must-do opportunity for any dedicated trilobite aficionado.

DISPLAYS

Once you've gathered together your trilobite collection and carefully curated your specimens, what do you do next? Perhaps you'll find an entertaining way to display your Paleozoic holdings. Some enthusiasts line shelves in their office with their trilobite treasures, and others proudly showcase them in well-lit living room cabinets. Either way, these incredible arthropods are guaranteed to be the subject of interest, scrutiny, and discussion wherever and whenever they appear.

NIOBE SP.

Lower Ordovician; Taous region; Erfoud, Morocco; 6.3 cm

(TOP, LEFT) *BELLACARTWRIGHTIA CF. JENNYAE* **LIEBERMAN AND KLOC, 1997**

Middle Devonian; Ludlowville Formation, Centerfield Limestone; Livingston County, New York, United States; 5.1 cm

(TOP, RIGHT) *CONOMICMACCA ALTA* **(LINAN AND GONZALO, 1986)**

Lower Cambrian, Bilbilian; Láncara Formation; Cantabrian Mountains, León Province, Spain; each trilobite 3.2 cm

(RIGHT) *ACHATELLA (VIRONIASPIS) KUCKERSIANA* **(SCHMIDT, 1881)**

Upper Ordovician, Caradoc Series; Kukruse Regional Stage; Alekseevka Quarry; St. Petersburg region, Russia; 3.1 cm

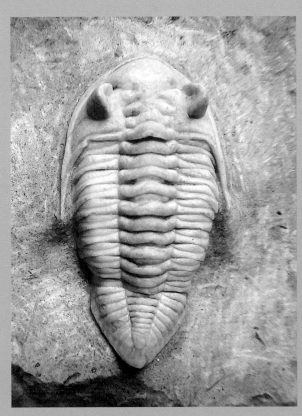

PSEUDOASAPHUS JANISCHEWSKYI BALASHOVA, 1976

Middle Ordovician, Lower Llanvirn Series; Asery Regional Stage, *Asaphus punctatus-cornutus* zone; Duboviki Formation; St. Petersburg Region, Russia; 10.1 cm

9

10 RENOWNED CARBONIFEROUS-PERMIAN TRILOBITE LOCALITIES

359 to 251 Million Years Ago

The Carboniferous Period (commonly separated into the Mississippian and Pennsylvanian in North American paleontological circles) began 359 million years ago following massive environmental changes and corresponding habitat destruction that marked the end of the Devonian Period. During this time, global temperatures rose dramatically, creating lush tropical swamps that pervaded the planet's equator-hugging landmasses. By now trilobites were barely hanging on in their chosen marine ecosystems, continuing their slow but steady descent toward evolutionary oblivion. Although this late-stage Paleozoic world may not have provided ideal conditions for trilobite continuity, it did supply a near-perfect environment for certain early plant species to flourish, which they did, eventually leaving behind the coal-rich deposits that provide the Carboniferous with its name. The oceans were also particularly warm, often forcing such diminutive trilobites as *Ameura major*, *Cummingella belisama*, and *Comptonaspis swallowi* (members of the Proetida order, which by the dawn of the Mississippian contained all remaining trilobite species) to burrow into the surrounding seafloor sediments to escape both the escalating water temperatures and an increasingly aggressive cast of predators, including early sharks and amphibians. By the beginning of the Permian, 299 million years ago, Earth's landmasses had once again all joined together to form the supercontinent called Pangea. The few remaining trilobite species that lived amid the coastal plateaus surrounding this huge landmass were uniformly small—under

WITRYIDES ROSMERTA G. HAHN, R. HAHN, AND BRAUCKMANN, 1986
Lower Carboniferous, Tournaisian (Lower Mississippian); Tournai Formation; Tournai, Belgium; 2.2 cm

6 centimeters in length. These diminutive arthropods played a relatively insignificant role in the planet's continuing faunal expansion—virtually all of which came to a crashing halt when the greatest mass extinction in Earth history wiped out 90 percent of life around the globe, subsequently bringing a dramatic close to both the Paleozoic Era and to the trilobites' 270-million-year swim through Deep Time.

The following reports describe 10 renowned trilobite-bearing Carboniferous and Permian deposits.

TOURNAI, BELGIUM

Located some 90 kilometers southwest of the ersatz European "capital" of Brussels, and sitting almost squarely on the border that separates Belgium and France, the land surrounding the historic town of Tournai has long served as the focal point for one of the richest Lower Carboniferous (Mississippian) faunas ever discovered. On the outskirts of this picturesque community dominated by towering church steeples and twelfth-century castles, dark-gray 350-million-year-old limestone deposits have yielded an intriguing variety of small (generally 3 centimeters or less) but significant proetid trilobites. These species include *Cummingella belisama*, *Witryides rosmerta*, *Paladin arduennesis*, and *Piltonia kuehnei*, virtually all of which are fossilized with a thin, alabaster-hued calcite coating. Although rock from the vicinity has been mined since Roman times, the area's quarries were first opened in the mid-nineteenth century to supply building materials for this fast-growing region that today supports more than two million inhabitants. More recently, many of these quarries have been shut down due to safety and economic concerns, making the trilobite material that has emerged from them in the last 170 years even more appreciated by both collectors and scientists.

ULUTAU MOUNTAINS, KAZAKHSTAN

As shown throughout these pages, evidence of fossilized trilobite remains can be found just about anywhere that properly aged sedimentary outcrops occur. Even high in the remote Ulutau Mountains—situated in the copper-rich Dzhezkazgan Region of Central Kazakhstan—complete 320-million-year-old examples of this ancient arthropod class have been discovered in impressive numbers. During the second decade of the twenty-first century, these relatively small Carboniferous (Pennsylvanian) trilobites, which average 3 centimeters in length, began appearing for sale, both on eBay and at prominent rock and

***DITOMOPYGE KUMPANI* (V. N. WEBER, 1933)**
Uppermost Pennsylvanian/Upper Carboniferous (Stephanian); Ulutau Mountains; Dzhezkazgan region; Karagandy Province, Kazakhstan; 2.7 cm

mineral shows throughout the world. Although often poorly preserved as internal molds on a beige-colored dolomitic limestone, some of these trilobites—which feature such closely aligned proetid species as *Ditomopyge kumpani* and *Griffithides praepermicus*—have been fossilized in conjunction with other marine fauna, including brachiopods and crinoid stems. Together these Paleozoic remnants provide a fascinating view of life in the primal seas during the closing stages of the trilobite line's tenacious trek through time.

CHOUTEAU FORMATION, MISSOURI

Following a planetwide extinction event at the end of the Devonian, the entire trilobite line found itself in dire evolutionary straits. During the preceding 170 million years, order after order of these intriguing invertebrates had vanished from the early oceans, leaving the entire class in a precipitous decline. At the dawn of the Carboniferous, only diminutive members of the Proetida order remained to live out the remaining 100 million years of the trilobites' lengthy crawl through the Paleozoic. Fossilized evidence of this last line of trilobites can be discovered in numerous spots across the face of the planet. The 350-million-year-old Mississippian-age Chouteau Formation outposts of Missouri, for example, have long been a favorite of North American collectors in search of small (1- to 3-centimeter-long) species such as *Ameropiltonia lauradanae* and *Comptonaspis swallowi*. These captivating creatures are noted for their finely detailed, three-dimensional preservation, a quality that has made some dedicated midwestern hobbyists focus almost exclusively on their acquisition. On occasion, fortunate Chouteau Formation field collectors have encountered groups of trilobites in close fossilized proximity. The preparation results of these discoveries often produce compelling multispecimen plates featuring up to half-a-dozen outstretched examples housed tightly together on a single sedimentary stone.

TIMOR, INDONESIA

By the time the Permian Period had shifted into high gear 298 million years ago, trilobites were barely hanging on in the evolving marine world. The average size of these amazingly adaptable arthropods had shrunk to only a few centimeters, and their speciation had reached a critical low. Despite these apparent trials and tribulations, the fossilized distribution of genera ranging from *Paraphilipsia* to *Acropyge* is rather impressive, with the often disarticulated remains of these terminal trilobites being found in sedimentary layers in such currently far-flung locations as Hungary, Indonesia, China, Pakistan, Russia, and Japan. Among these locales, a special focus should be placed on the Maubisse Formation of Timor, Indonesia, which has provided a number of intriguing complete (although usually enrolled) examples of *Neoproetus indicus*, *Hildaphillipsia hildae*, and *Phililipsia parvula*—most no larger than a small pea and all preserved in an ivory-to-pink-toned calcite. The mere existence of these diminutive denizens of the deep (well, to be honest, shallow shelf environments) proves that trilobites were still contributing a significant "something" to their class's noble heritage despite being on their last multijointed legs. The various Permian formations that appear throughout the Timor vicinity have proven to be among the richest on the planet, with more than 600 faunal species—including corals, brachiopods, and crinoids—emphatically showing that trilobites were far from alone in their steadily changing undersea kingdom.

CABALLERO AND LAKE VALLEY FORMATIONS, NEW MEXICO

During the last two decades, New Mexico has emerged as a Mississippian Period trilobite epicenter, with an impressive 23 species of these ancient arthropods identified from the neighboring Caballero and Lake Valley formations, including the likes of *Piltonia carlakertisae*, *Namuropyge newmexicoensis*, and *Pudoproetus fernglenensis*. At times these trilobites have initially been identified from the multitude of disarticulated fragments—usually well-preserved pygidia—that litter their sedimentary exposures. But when found complete, these distinctive proetids can reach 5 centimeters in length, quite large in comparison to most other Carboniferous specimens. These trilobites are frequently preserved in a dark-brown or black calcite,

and when properly prepared this contrasts vividly against the Lake Valley Formation's characteristic grayish-pink matrix. This attractive coloration, as well as their increasing scarcity on the commercial market, has made these southwestern species popular additions to many private collections. Unfortunately, as is the case with other once flourishing trilobite sites around the globe, recently enacted legislation has made the retrieval of these unusual Paleozoic remnants increasingly difficult, even for those actively involved in academic pursuits.

LODGEPOLE FORMATION, MONTANA

No matter how many times any of us may attempt to recognize and rationalize the seismic changes that have continually transformed Earth's surface over the last half-billion years, as one drives though the mountainous environs of southern Montana, it's still difficult to fathom that this high-altitude landscape was once located at the bottom of a semitropical sea. In the Lower Mississippian, some 350 million year ago, this imposing region that now comprises Gallatin County was part of a rich marine shelf environment. That ecosystem apparently proved most hospitable to a host of diminutive trilobite genera such as *Brachymetopus* and *Australosutura*, each representing the survivors of a Devonian-ending mass extinction event that wiped out more than 60 percent of life on Earth. Today the fossils of these late-stage trilobites are found at a significant elevation in the tree-lined Centennial mountain range. Merely reaching any of the 2,500-meter-high dig sites is a yeoman's task—as is extracting these small (generally under 3-centimeter-long) trilobites from their hard limestone encasements. Recovered specimens tend to be beautifully preserved in a thick, golden-brown calcite, and many of the key morphological features of their exoskeletons are displayed in fine detail.

***NEOPROETUS INDICUS* TESCH, 1923**

Lower Permian; Maubisse Formation; West Timor, Indonesia; 3.5 cm

VOLGOGRAD, RUSSIA

Even for the more perceptive among us, it's sometimes challenging to comprehend the sheer vastness of the Russian landscape. More than 70 times the size of the United Kingdom, 5 times the size of India, and almost twice the size of the United States, it's not surprising to learn that Paleozoic deposits dating from the Cambrian all the way to the Permian dot this enormous tract of terra firma—and many of those outcrops contain trilobites of various sizes, shapes, and species. As tantalizing as digging for these primordial organisms may initially appear to be, merely reaching many of these Russian trilobite strongholds—including the Pennsylvanian-age Moscovian Regional Stage (source of *Ditomopyge producta*) and the Permian-age Arti Formation (home to *Ditomopyge artinskiensis*)—requires more than the simple desire to do so. Arranging an excursion to some of Russia's less accessible Carboniferous and Permian sites can present enough logistical headaches to make any right-minded soul second guess their original intent. Such distance-defying treks take persistence, patience, and planning, along with an appetite for tackling some of the most formidable obstacles nature has at its disposal. When you finally approach any of these fossil hotspots, including those near the city of Volgograd—some 800 kilometers southeast of Moscow—you'll encounter challenging roads and temperatures designed to test the limits of human endurance.

WUPPERTAL, GERMANY

Despite being collected on and off for decades, it has only been during the initial years of the twenty-first century (which coincidentally corresponds with the full-fledged emergence of the internet) that the extensive trilobite reserves of Germany have become fully appreciated by arthropod enthusiasts around the globe. The nation's diverse Devonian trilobite localities, especially those hailing from the Eifel Region of the Rhine Valley, have continually drawn both academic and hobbyist interest (mostly due to their excellent, calcite-covered preservation and impressive variety of species). But the Lower Carboniferous trilobites of Germany—almost all of which emerge from deposits found near the midsized city of Wuppertal in the nation's western region—have only recently piqued a similar degree of worldwide collector curiosity. These 330-million-year-old examples include such closely related species as *Waribole laticampa*, *Spinibole coddonensis*, *Archegonus aprathensis*, and *Archegonus nehdenensis*, many of which are found complete and often reach lengths of 3 to 4 centimeters. For the most part, these latter-stage proetids have been fossilized as lightly mineralized internal molds. This form of preservation serves to effectively capture the streamlined yet simplistic body design that distinguished trilobite morphology as the class slowly approached the end line of its quarter-billion-year crawl through the Paleozoic seas.

CRAWFORDSVILLE, INDIANA

Although far more renowned for their world-famous crinoid assemblages (featuring more than 60 recognized species), the Lower Mississippian Edwardsville Formation quarries located near the rural Indiana town of Crawfordsville have also yielded impressively sized *Hesslerides bufo* trilobites, some 5 centimeters in length. On rare occasions, these two faunal elements emerge in close association on the quarry's 340-million-year-old limestone slabs, creating a dramatic captured in time look at life in the primordial seas. Ironically, those who specifically seek the magnificent three-dimensionally preserved crinoids, for which this locale has been justifiably lionized since its discovery in 1842, usually show little to no interest in any trilobite matter that may appear on their sedimentary stones. A few collectors have even admitted to having the quarry's trilobite detritus—especially if it's an incomplete specimen—surgically removed from their finds to best preserve the stoic sanctity of their otherwise unsullied crinoid discoveries. Of course, most who stumble on a collection-worthy *Hesslerides* are only too eager to either save it, show it off, or sell it to someone who may more fully appreciate its importance as one of the last trilobite species to ever swim through the planet's murky seas.

DURHAM, ENGLAND

For a nation contained within relatively limited geographical boundaries, the British Isles are packed with some of the planet's most renowned, prolific, and studied trilobite repositories. Among those diverse fossiliferous formations are the rugged Middle Cambrian cliffs of St. David's, Wales (*Paradoxides davidis)*, the challenging Ordovician fossil "pockets" of Girvan, Scotland (*Paracybeloides girvanensis*), the famed Silurian Much Wenlock deposits of Dudley, England (*Calymene blumenbachii*), and the perhaps somewhat lesser known Carboniferous outcrops that appear near the cathedral-filled city of Durham, England. Over the centuries, each of these legendary sites has been the subject of acute scientific scrutiny, as well as the focus of intense collector interest. Yet the Carboniferous trilobites of Great Britain remain something of an underappreciated resource, primarily because many of these late-stage proetids look so strikingly similar. However, amid this sea of conformity, certain Durham genera do manage to stick out, both figuratively and literally, perhaps most notably *Weberides sp.*—a trilobite with a distinctively pointed pygidium. This generally diminutive species (averaging between 2 and 3 centimeters in length) can be found either enrolled or outstretched. It has been uncovered from similarly aged sediments in both Ireland and Scotland, as well as within its most recognized Paleozoic portal, which occurs along the River Wear in the northeast of England.

(LEFT) *ARCHEGONUS (PHILLIBOLE) NEHDENENSIS* HAHN AND HAHN, 1969

Lower Carboniferous (Mississippian), Viséan Stage; Kulm cu II/III; Aprath, Wuppertal, Germany; 3.1 cm

(BOTTOM) *DITOMOPYGE SP.*

Middle Pennsylvanian; Bond Formation; LaSalle Limestone; Pontiac, Illinois, United States; 2.1 cm

(TOP) *GRIFFITHIDELLA DORIS* (HALL, 1860)

Lower Mississippian; Lake Valley Formation; Caballo Mountains, New Mexico, United States; 1.8 cm

(LEFT) *WEBERIDES SP.*

Upper Devonian/Lower Carboniferous border; Topmost Famennian Stage; Devon, England; 3.2 cm

(LEFT) *AMEURA SANGAMONENSIS* **(MEEK &
WORTHEN, 1865)**

Upper Pennsylvanian; Lonsdale Limestone; Wildlife Prairie
Park; Peoria, IL, United States; 3.0 cm

(BOTTOM) *DITOMOPYGE ARTINSKIENSIS*
(V.N. WEBER, 1933)

Lower Permian, Artinskian; Arti Formation; Krasnoufimsk;
Middle Ural, Russia; 4.4 cm

10 KEY CURATION DETAILS

Let's say you've managed to assemble a sizable trilobite collection containing scores—if not hundreds or even thousands—of specimens from all corners of the globe and all aspects of Deep Time. They line cabinets in your home, fill shelves in your office, and overflow boxes in your upstairs closet. Perhaps the moment has come to make better sense of your voluminous holdings by properly curating your collection. Thanks to the prodigious amount of information available on the internet—ranging from enlightening museum-sponsored sites (where a variety or trilobite-oriented papers, photos, and journals can be found) to amazingly detailed amateur pages—every collector has access to a world-class compendium of essential trilo-data literally at their fingertips. The simple fact is that no matter its place of origin, its age, its rarity, its condition, or the degree of academic attention it may have already garnered, every collection-worthy trilobite should be provided with an accompanying identification label and number, one that can be quickly and easily cross-referenced with a list stored conveniently in a notebook, on a series of index cards, or in a well-marked computer file. For a multitude of trilobite devotees, the proper curation of their collection ranks among the most enjoyable aspects of their hobby. In many cases, this process provides these enthusiasts with a true sense of accomplishment, a realization that due to their efforts some 500 million years after that particular trilobite met its untimely demise it now enjoys a restored level of identity—one that will enable it to be studied, admired, and enjoyed for many eons to come.

OLENELLUS TERMINATUS (PALMER, 1998)
Lower Cambrian, Series 2, Dyeran; Pioche Shale, Comet Member;
Ruin Wash; Pioche, Nevada, United States; 4.2 cm

Here are 10 key curation details that you should include on any trilobite identification label, explained here using the information for *Olenellus terminatus.*

GENUS

The genus is the taxonomic category that falls between family and species. For a trilobite such as *Olenellus terminatus, Olenellus* represents its generic classification.

SPECIES

Using the same trilobite example, *terminatus* represents the species. Many trilobites of the genus *Olenellus* exist around the globe, but *O. terminatus* hails exclusively from outcrops found across the western United States.

AUTHOR

The author references the academician who first described a certain trilobite in scientific literature. In our ongoing example, the label would read *Olenellus terminatus* (Palmer, 1998), which gives proper credit to the noted paleontologist Pete Palmer.

YEAR

Not surprising, the year in which the author first published their trilobite findings in an appropriate academic publication is the year referenced here. In our example, the year of publication for Palmer's description of *Olenellus terminatus* was 1998.

GEOLOGIC AGE

The Paleozoic Era lasted for 290 million years, and trilobites, in one form or another, survived for nearly the entirety of that extended stretch of planetary history. *Olenellus terminatus* appeared quite early in the trilobites' crawl through Deep Time, emerging in the Lower Cambrian, some 518 million years ago.

FORMATION

One of the most important snippets of information that a properly prepared label can provide is the geologic formation from which the trilobite was recovered. In the case of *Olenellus terminatus* that formation was the Pioche Shale.

MEMBER

Although not always acknowledged on even the most descriptive labels, it is helpful to know the geologic member; that is, the larger sedimentary unit in which a particular formation can be found. In our *Olenellus* scenario, the Pioche Shale is a Comet Member.

LOCATION

Often a label's location will denote the quarry, city, province, or state in which a trilobite species is found. In our example, *Olenellus terminatus* was discovered in the Pioche Shale, which is basically in the middle of nowhere, so the location recorded is Ruin Wash, Nevada.

COUNTRY

This detail is somewhat self-explanatory; it is the nation within whose borders a specimen was discovered. For our *Olenellus terminatus*, that country is the United States.

NUMBER

Although this is not a detail included by all collectors, many enthusiasts find it helpful to place a small catalog number on each specimen, so it can be quickly cross-referenced against a master list stored on a computer or in a notebook.

(TOP) **PSEUDOCALYMENE SZECHUANENSIS (LU, 1962) AND CHEIRURUS N. SP.**

Lower Ordovician, Upper Tremadocian; Fengxiang Formation;
Liexi, Hunan, China; largest trilobite 5 cm

(RIGHT) **DROTOPS MEGALOMANICUS MEGALOMANICUS STRUVE, 1990**

Middle Devonian, Basal Givetian; Bou Dîb Formation, Lower
Member; Jebel Issoumour; Maïder region (near Alnif), Morocco;
14 cm

***STELLA DEMISSA* EGOROVA AND SAVITSKY, 1968**

Middle Cambrian; Hatanga Formation; Anabar River, Northern Siberia; 8.1 cm

***ALBERTELLA LONGWELLI* (PALMER AND HALLEY, 1979)**

Middle Cambrian; Carrara Formation; Pahrump, Nevada, United States; 7.9 cm

11

10 OBSCURE (BUT STILL SIGNIFICANT) TRILOBITE LOCALITIES

W hen work began on *The Trilobite Collector's Guide* a few years ago, the paramount, clear-cut, nothing else matters editorial objective was to place an unwavering fossiliferous focus on the most renowned, important, and studied trilobite repositories in the world. During that process, scores of globe-spanning Paleozoic sites have been examined in varying degrees of descriptive and paleontological detail. These reports range from tales of the legendary Burgess Shale in western Canada to accounts dedicated to the lesser-known but equally intriguing Fezouata deposits of Morocco. Despite my best attempts to present a comprehensive, collector-friendly compendium of Earth's leading trilobite localities—whether found amid the islands off South Australia, atop the mountains of southern California, or adjoining the hills of southern China—some noteworthy but rather obscure sites run the risk of slipping through the sedimentary cracks. These often overlooked fossil-filled formations range in their geographic distribution from the rarely explored Lower Cambrian outcrops of Sardinia, Italy, to the frequently forsaken Middle Carboniferous layers of Wise County, Texas.

Here is a salute to 10 additional Paleozoic localities that have occasionally found themselves on the cutting edge of scientific research and at the epicenter of collector fascination despite their less than stellar renown.

OLENELLUS ROMENSIS RESSER AND HOWELL, 1938

Lower Cambrian; Rome Formation, Montevallo Shale; Montevallo, Alabama, United States; 6.9 cm

CAMBRIAN PERIOD

ROME FORMATION, ALABAMA, UNITED STATES

There has been a long-standing and well-established academic belief that trilobites were strictly and solely marine inhabitants. But in 2014, scientists investigating a series of Lower Cambrian deposits in the Appalachian Mountains of Tennessee discovered something they had not anticipated. In outcrops of the 515-million-year-old Rome Formation, a team led by researchers from the University of Saskatchewan found unmistakable fossil evidence indicating that some olenellid trilobites—including those belonging to the formation's dominant species, *Olenellus romensis*—had already briefly journeyed onto sea-adjacent tidal flats, possibly to feed or to lay their eggs, soon after beginning their lengthy passage through the Paleozoic Era. This formation can also be seen in Alabama, where complete *Olenellus* and *Mesonacis* specimens have been uncovered, although none so far with the accompanying *Cruziana* (marks made by trilobite walking legs) that would provide evidence of sojourns out of their native sea habitat.

NEBIDA FORMATION, SARDINIA, ITALY

Most travelers who make their way to Sardinia are there for the fun, the sun, and the food. But in certain paleontological circles, this large Italian island situated in the middle of the Mediterranean Sea is far more renowned for Lower Cambrian fossil deposits that produce such rare trilobites as *Dolerolenus zoppii* and *Sardaspis laticeps*. A dozen trilobite species have been described from the fossil-bearing outcrops found throughout the island's southwestern corner, most in the thick limestone layers of the Nebida Formation. Equally notable is that some of the vicinity's trilobite

genera—including a species of *Eoredlichia*—also occur in similarly aged strata in southwest China, and others, including *Dolerolenus*, have been uncovered in Cambrian horizons known from the Iberian Peninsula.

BUEN FORMATION, SIRIUS PASSET, GREENLAND

Of all the distant, isolated, and hard-to-reach destinations where trilobite fossils have been discovered, perhaps none can rival the sheer seclusion presented by the Sirius Passet Lagerstatte. This Lower Cambrian outcrop borders the Arctic Sea along the northernmost coast of Greenland, some 1,600 kilometers above the Arctic Circle. Initially brought to international attention in 1984 by explorers working in conjunction with that nation's geological survey—and named in honor of the area's renowned Sirius Dog Sled Patrol—six small outcrops of the 518-million-year-old Buen Formation were revealed among the sedimentary exposures that run adjacent to the ice-filled J. P. Koch Fjord. In addition to a variety of sponges, brachiopods, and soft-bodied arthropods, the fauna presents a small selection of trilobite species, including *Buenellus higginsi*, which represent some of the earliest known examples of their noble arthropod line. Preservation at these hard-to-reach sites has been somewhat problematic—with the fossils generally appearing as poorly preserved elements in large microbial mats—but the Sirius Passet locale holds the promise of revealing some startling discoveries as additional fieldwork is conducted in this remote corner of the globe.

MAYAN STAGE, ANABAR PLATEAU, SIBERIA

A variety of Middle Cambrian trilobite genera are represented in the fossiliferous layers that emerge in Siberia's Sakha Republic. These examples can best be uncovered in the remote sedimentary deposits provided by the 510-million-year-old Mayan Stage of the Anabar Plateau—a rugged spot of permafrost-covered tundra located high above the Arctic Circle, some 40 kilometers west of the area's imposing Lena River. Despite its remote location, the Anabar trilobite material has been surprisingly abundant. Enough complete specimens of species such as *Hatangia scita*, *Michaspis librata*, and *Urjungaspis picta* have been found that these small 2- to 4-centimeter-long arthropods are now readily available on the commercial market. More than a dozen different trilobite species have been unearthed from the plateau's various Mayan Stage outcrops, and the diversity of morphological configurations seen in these hard limestone sediments indicates just how rapidly—at least in a geological sense—the trilobite class was evolving at this nascent period in its history.

ORDOVICIAN PERIOD

COROICO FORMATION, CAMARGO, BOLIVIA

One of the most intriguing and out-of-the-way Paleozoic localities ever to yield a scientifically significant bounty of trilobite material can be found at a breathtaking altitude of more than 2,400 meters, high along the southern ridge of the Andes Mountains. There, in the outskirts of Camargo, Bolivia, a prolific outcrop of the Coroico Formation has yielded a veritable bumper crop of fascinating Ordovician-age fossils, including such widely recognized trilobites as *Hoekaspis matacensis* and *Hypermecaspis branisai*. These unusual species frequently emerge from their 430-million-year-old burial grounds in large (up to 15-centimeter-long) limestone concretions. When these hard, pancake-shaped nodules are carefully

cracked open, they may reveal fossils with positive/negative splits that display outstanding preservation and detail.

VIIVIKONNA FORMATION, ESTONIA

The neighboring fossil-packed Ordovician outcrops of western Russia long ago gained their warranted degree of academic acclaim and collector kudos. But in recent years the similarly aged Viivikonna Formation of Estonia has become equally recognized for yielding an impressive array of well-preserved 450-million-year-old trilobite treasures. These Russian and Estonian outcrops share a variety of faunal elements, but the material emerging from the Estonian deposits near the town of Narva—including trilobite species such as *Apianurus kuckersianus*, *Chasmops tallinnensis*, and *Estoniops exilis*—tends to be smaller in size and appears on a brittle, brick-colored sedimentary matrix. That highly friable stone is often covered with scattered fossiliferous debris reflecting the remains of the surrounding seafloor life, which includes impressive examples of bryozoans and crinoids. When this attractive material began to emerge on the international market in 2018, it almost immediately became the subject of intense collector focus.

ANTELOPE VALLEY FORMATION, NEVADA, UNITED STATES

The Antelope Valley Formation is situated almost squarely on the state line that divides Nevada and California. At times it's hard to know exactly where one stands amid the wide-open plains and dry desert gulches that distinguish the terrain surrounding this rarely visited Paleozoic repository. The most noteworthy fossiliferous outcrop of this limestone escarpment occurs in the shadows of Meiklejohn Peak (near the thriving Nevada town of Beatty, population 999) where such seldom seen species as *Lachnostoma latucelsum* and *Pseudomera barrandei* can occasionally be found as complete, articulated specimens—some more than 10 centimeters in length. The less than clear-cut boundaries that mark the area's fossil-filled formations lend themselves to ongoing scientific discussion and debate. As more study is done and more discoveries are made, it is hoped that both the borders and the contents of the Antelope Valley Formation will eventually come into better focus for academics and amateur enthusiasts.

DEVONIAN PERIOD

BOIS BLANC FORMATION, ONTARIO, CANADA

The sheer immensity of the Canadian landscape is responsible for the less than stellar status enjoyed by some of the previously unknown, unexplored, and unimagined trilobite localities found throughout the Great White North. From the rugged cliffs of Quebec's Anticosti Island in the east to the snow-capped peaks of the Rosella Formation in the west, Canada is a veritable coast-to-coast

DOLEROLENUS ZOPPII (MENEGHINI, 1882)
Lower Cambrian, Series 2; Nebida Formation, Punta Manna Member; Porto di Canal Grande, Sardinia, Italy; 3.2 cm

Paleozoic playground. One of the nation's more obscure Devonian outcrops, the Bois Blanc Formation, occurs along the U.S.-Canada border in southern Ontario. Within the formation's distinctive beige-toned limestone layers, tantalizing hints of one of invertebrate paleontology's most legendary creatures—the giant lichid trilobite *Terataspis grandis*—have been recovered in small but significant numbers. An almost complete example (now housed at the Royal Ontario Museum) was unearthed in the Bois Blanc in 2010, and rumors of other semicomplete specimens continue to rumble through the trilobite grapevine. If that wasn't enough to titillate the trilobite-craving soul of any serious collector, the enticing remains of equally rare species—including undescribed cheirurids and phacopids—are also present throughout this site.

VOORSTEHOEK FORMATION, CERES, SOUTH AFRICA

Whenever African trilobites are discussed in fossil-centric circles, invariably it is the dizzying diversity of material emerging from the bountiful deposits of Morocco that are the primary and often sole subject of conversation. But other notable Paleozoic horizons also exist on this vast and often mysterious continental mass. About 7,500 kilometers southeast of Casablanca, near the town of Ceres, South Africa, lies the infrequently explored Voorstehoek Formation. There, amid Devonian-age outcrops of the Bokkeveld Group, a combination of well-defined mudstone and siltstone layers have yielded an intriguing but rarely studied fauna of well-preserved trilobite material. The most common species found in this South African repository—although still unusual, especially in complete form—is *Gamonedaspis boehmi*, a dalmanitid strikingly akin to a trilobite genus now uncovered in similarly aged outcrops throughout Argentina.

CARBONIFEROUS AND PERMIAN PERIODS

GRAFORD FORMATION, TEXAS, UNITED STATES

The old saying goes that everything grows bigger in Texas. Perhaps that's true, but in all honesty the Lone Star state's trilobites tend to run a tad on the scrawny side—even for specimens measured on the generally undersized Late Paleozoic size scale. But no one should blame Texas for producing the diminutive trilobites (average length of 3 centimeters) that some 300 million years ago inhabited the seas that then covered this now often parched and always imposing land. By the beginning of the Pennsylvanian Period, only trilobites of the proetid order had managed to survive the outrageous fortune of evolutionary change, and well preserved examples of *Ditomopyge scitula* can be found littering the ground throughout the Graford Formation of Wise County. These brick-red remnants are usually recovered as disarticulated fragments, but complete specimens can occasionally be collected while strolling through the limestone deposits near the town of Bridgeport, about an hour's drive northeast of beautiful downtown Dallas.

(LEFT) *PARANEPHROLENELLUS KLONDIKENSIS* WEBSTER, 2007

Lower Cambrian, Dyeran; Pioche Formation, Delamar Member; Klondike Gap, Lincoln County, Nevada, United States; Each trilobite 3.5 cm

(BOTTOM) *HYPERMECASPIS BRANISAI* PŘIBYL AND VANĚK, 1980

Middle Ordovician; Coroico Formation, Hoekaspis yahuari Biozone; Camargo, Chuquisaca Department, Bolivia; 8.3 cm

(TOP) *APIANURUS KUCKERSIANUS* (SCHMIDT, 1885)

Upper Ordovician; Kukruse Regional Stage; Viivikonna Formation, Kiviõli Member; Narva, Northeast Estonia; 2.5 cm

(BOTTOM) *DIACALYMENE SP.*

Middle Silurian; Thorneloe Formation; Lake Timiskaming mining district; Northern Ontario, Canada; 5.2 cm

(BOTTOM, RIGHT) *GAMONEDASPIS BOEHMI* (KNOD, 1908)

Middle Devonian, Lowest Eifelilan; Voorstehoek Formation; Ceres, Western Cape, South Africa; 5.4 cm

12

10 CREATURES THAT CAME BEFORE TRILOBITES

From a contemporary perspective, trilobites seem to have emerged fully formed upon the Cambrian scene. By the time the first members of this ancient line of arthropods began inhabiting marine environments 521 million years ago, they were already creatures with highly developed eyes, complex digestive systems, and admirably functional calcite carapaces. Clearly, at that moment in evolutionary time, trilobites represented the most advanced life-forms ever produced on Earth. Due to their sudden and dramatic appearance in Paleozoic oceans, it is apparent that trilobites evolved from earlier, more primitive organisms—creatures that left minimal evidence in the fossil record of their long-ago existence. Some of these ancestral forms may have resembled segmented worms or articulated jellyfish more than the distinctive three-lobed animals we all recognize today. At least a few of those possible Precambrian trilobite ancestors—such as *Spriggina flounderosi*, found in the 550-million-year-old Edi-acaran-age rocks of Australia—appear to have possessed rudimentary body segments and even basic genal spines. Despite these recognizable anatomical advances, these primeval organisms were a far cry from the trilobites of the Redlichiida order that some 30 million years later would dominate the world's seas. The Lower Cambrian development of key morphological features—such as a resilient outer shell, itself indicative of the trilobites' ascension as major players in the drama surrounding complex life's first tentative steps on our planet—were most likely the arthropod response to ever-growing threats in their marine habitat, which included everything from the emergence of more menacing predators to increasingly unpredictable climatic conditions.

MESONACIS SP.
Lower Cambrian; Kinzers Formation; Lancaster, Pennsylvania, United States; 5.1 cm

Note the primitive, elongated shape of this early genus.

Here is a look at 10 creatures that lived before trilobites, along with a few concurrent Paleozoic arthropods, that provide some much-needed insight into trilobite history and morphology.

SPRIGGINA FLOUNDEROSI

Spriggina flounderosi holds the distinction of being the "official fossil emblem" of South Australia. Yet when examples of these 3- to 5-centimeter-long, bilaterally symmetrical animals were first identified in that region's 550-million-year-old Ediacaran-age rocks, the paleontological community faced a major problem in determining a proper taxonomic

category in which to place them. With their rudimentary body segments, basic genal spines, and rows of interlocking dorsal plates, in the years following their discovery in the 1950s, *Spriggina* has been classified as everything from a primitive annelid worm to an early arthropod—the latter indicating that they may, in fact, be among the possible forerunners of trilobites.

BOMAKELLIA KELLERI

The White Sea and Syuzma River regions of northern Russia have become renowned in paleontological circles for their preservation of 555-million-year-old soft-bodied fossils. These specimens include the strange, leaflike creature *Bomakellia kelleri*, which some scientists believe may have been the first organism on the planet capable of possessing sight. Only a single 9-centimeter-long example of this unusual species has been found in the vicinity's Ediacaran-age rocks. It was initially classified in the problematic arthropod class Paratrilobita, but *Bomakellia* has since been reclassified into the little understood class Rangeomorph, although that categorization remains very much subject to change as more information and more specimens become available for study.

PARVANCORINA MINCHAMI

Living along the Precambrian seafloor some 550 million years ago, the diminutive shieldlike creature known as *Parvancorina minchami* has been noted for its superficial morphological similarities to some Lower Cambrian trilobite species—and it is frequently mentioned as a possible trilobite precursor. First discovered in the mid-1950s in the Ediacaran-age Flinders Range of South Australia, in the following decades similar primal deposits in the White Sea area of Russia have yielded a significant sample of these 1- to 2-centimeter-long creatures, which often appear in multispecimen assemblages on the same rock surface. Their strange, rather fanciful bow-and-arrow shape has made some scientists attribute *Parvancorina* to the arthropod line, but their apparent inability to walk—or move in any self-propelled manner—serves to minimize any such manifest arthropod affinity.

ARCHAEASPINUS FEDONKINI

Best known from a smattering of examples found in the Precambrian-age Penega Formation outcrops of East Angelsk, White Sea, Russia, *Archaeaspinus* ranks among the least understood organisms recognized from this often confusing period of Earth history. Currently attributed to the phylum Proarticulata due to its early demonstration of bilateral symmetry, *Archaeaspinus* has occasionally been referenced as a contender in the category of possible trilobite ancestor. From appearance alone, however, it seems that evolution would have needed to dig deep into its endless bag of tricks to transform this simplistically designed organism into an early Redlichid trilobite—even if such a process could have occurred over a 30-million-year span.

SKANIA FRAGILIS

The mere notion that *Skania fragilis* is a Middle Cambrian creature—an arthropod that existed some 15 million years *after* the first trilobites emerged in the seas—would seem to logically and effectively end any speculation that it may have played a role in the ascendancy of the trilobite class. And, quite frankly, it does. But its striking resemblance and possible relationship to the Precambrian creature *Parvancorina*—an organism that ranks near the top of many scientific lists when it comes to possible trilobite predecessors—makes it reasonable to include it here. Only one complete example of *Skania* has been discovered so far, and it was found in the legendary Burgess Shale layers of British Columbia.

PRIMICARIS LARVAFORMIS

Since its scientific discovery in 1984, China's Lower Cambrian Chengjiang Biota has proven to be one of the supreme repositories of early life on this planet. Among the scores of previously unknown arthropod species revealed in the formation's bands of thick, mustard-toned mudstone is the small creature *Primicaris larvaformis*—once thought to be (as its name implies) the larval or meraspid stage of the equally enigmatic early arthropod *Naraoia*. Only a single specimen of *Primicaris* has been unearthed so far, and some scientists view it as the possible descendant of a Precambrian line that may have also led to the emergence of trilobites.

KUAMAIA LATA

Among a spate of presumed early trilobite relatives, *Kumaia lata* represents one of the more intriguing. A contemporary of Lower Cambrian trilobites, this complex, midsized arthropod (examples have been found up to 10 centimeters in length) has been classified as part of the recently created taxonomic grouping known as artiopoda, which also includes trilobites. Due to its impressive size and sophisticated morphology (including 15 pairs of jointed legs and an intricate system of gills), it is speculated that *Kuamaia* may have been a mobile hunter and scavenger that continually scoured the primal seafloor in search of expeditious feeding opportunities.

"TRILOBITOMORPHS"

When the hallowed *Treatise of Invertebrate Paleontology* (Part 0: Vol. 1) was released in 1959, trilobitomorph was effectively utilized as a comprehensive term designed to encapsulate virtually every non-biomineralized arthropod, including the Cambrian likes of *Marrella splendens*, known from locations such as the renowned Burgess Shale. Since then scientific insight and knowledge have continued to escalate in direct proportion to the number of soft-tissue preserving fossil sites found worldwide. In accordance, our understanding of the various taxonomic families into which these soft-bodied creatures should be properly assigned (including Ordovician marrellomorph genera such as *Furca*) has also continued to expand. Today we recognize the term *trilobitomorph* as an antiquated although strangely endearing term that represents a bygone era in paleontological history.

NARAOIA SP.

A soft-bodied arthropod perhaps best known from the Lower Cambrian Chengjiang outcrops of southern China, complete examples of *Naraoia* have also been uncovered in both the Middle Cambrian Burgess Shale of western Canada and the similarly aged Wheeler Shale of central Utah. Due to their primitive yet intriguing morphological profile, these small (generally under 3-centimeters-long)

ovate organisms were long thought to represent the taxonomic line that may have produced trilobites. Some late-twentieth-century scientists even classified *Naraoia* in the trilobite class, but others insist that they are more aligned with the crustacean phylum that includes such animals as isopods and shrimp.

SOFT-BODIED TRILOBITES

The now outdated catch-all classification of "soft-bodied trilobites" was initially designed to aid mid-twentieth-century scientists in their daunting quest to classify the dizzying array of nonbiomineralized creatures they had found. These organisms possessed neither the protective outer shells of their trilobite cousins nor the size, strength, or smarts required to ensure their continued existence, so the life of these primitive, sea-dwelling inhabitants may well have been short and not particularly sweet. One can imagine the tenuous times for these unimposing arthropods, playing hide-and-seek with predators as they darted in and out of ancestral sponge beds and under sturdy stromatolite reefs in their attempts simply to make it through one more day.

(LEFT) *MODOCIA KOHLI* ROBISON AND BABCOCK, 2011

Middle Cambrian, Series 3, Drumian; Middle Wheeler Formation; House Range and Drum Mountains; Utah, United States; 3.2 cm

(BOTTOM) *KUAMAIA LATA* HOU, 1987

Lower Cambrian; Maotianshan Shales; Heilinpu Formation; Chengjiang, Yunnan Province, China; 9.7 cm

OLENELLUS SP.

Lower Cambrian; Rome Formation, Montevallo Shale; Montevallo, Alabama, United States; 6.9 cm

FURCA SP. **(MARRELLOMORPH)**

Lower Ordovician, Arenigian; Upper Fezouata Shales; Zagora, Morocco; 15.2 cm

OLENELLUS CORDILLERAE

Lower Cambrian; Carrara Formation; Echo Shale; Resting Springs, California, United States; 6 cm

13

10 EMINENTLY ELEGANT TRILOBITES

Even among those who consider themselves top-tier invertebrate aficionados, "elegant" is a word rarely used when describing members of the calcite-covered trilobite class. Generally, when the subject turns to these ancient arthropods, collectors—as well as those in their immediate circle of friends and family—either appreciate the inherent, time-tested grandeur of these renowned "butterflies of the sea" or regard them as little more than fossilized sea slugs. Yet, for many trilobite enthusiasts, the elegantly eccentric configurations presented by certain prominent (and not so prominent) trilo-types serve to enhance both their scientific significance and their primal appeal. In the eyes of a notable number of hobbyists, these organisms have come to represent some of early life's most captivating and collectible efforts due to their often outlandish appearance and the incredible duration of their passage through Earth's history. From the graceful forms displayed by 520-million-year-old Cambrian olenellids to the streamlined shapes showcased by 320-million-year-old Carboniferous proetids, trilobites exhibited a previously unprecedented degree of natural nautical charm. Indeed, for many Paleozoic devotees around the globe, it is their often quirky elegance that makes collecting trilobites so eternally entertaining.

What follows is a highly subjective analysis of 10 of the most eminently elegant trilobites ever to emerge from the planet's Paleozoic seas.

QUADROPS FLEXUOSA (MORZADEC, 2001)

Middle Devonian; El Oftal Formation; Jebel Ofaténe, Morocco; 7.5 cm

QUADROPS FLEXUOSA

Resembling a top-billed attraction from the latest *Alien* movie, or at least an unfriendly visitor from another world, this Devonian species, found in the fossil-filled sedimentary layers of Morocco, is an amazing example of the strange and unpredictable tricks that evolution can play. There has been some speculation that the multitude of spines adorning this 7-centimeter-long creature's calcite carapace may have served as sensory organs alerting the trilobite to possible enemies or even to prey. Or those plentiful pointy barbs may have simply been what they appear to be—highly developed defensive weapons that helped *Quadrops* survive in an increasingly competitive primal environment.

REMOPLEURIDES ASTEROIDEUS

The mind-boggling morphological diversity inherent in the trilobite class is one of the elements most appealing to collectors across the planet. That diversity clearly manifests itself in members of the *Remopleurides* genus—few trilobite genera are more elegant, more exotic, or more superficially strange. With its huge crescent-shaped eyes, *R. asteroideus* was most likely a pelagic species, spending much of its life navigating through sun-dappled, shallow water environments where its sleek body design and superior eyesight served as major advantages when it came to both spotting prospective prey and avoiding possible predators.

GABRICERAURUS DENTATUS

Not every trilobite requires multiple rows of freestanding spines or an overly detailed morphological design to justify its inclusion on this "most elegant" list. Some fossilized examples may simply represent the quintessence of anatomical grace within the trilobite kingdom. Perhaps no species conveys that criterion with more Paleozoic aplomb than the appealing *Gabriceraurus dentatus*, one of the largest cheirurid trilobites (reaching 12 centimeters in length) known from the often explored Ordovician outcrops of Ontario, Canada.

DICRANOPELTIS SCABRA

With its distinctive, brightly colored mineralized carapace offering a dramatic contrast to the Motol Formation's characteristic chocolate-brown mudstone matrix, the rare Czech species *Dicranopeltis scabra* certainly ranks among the paleontological realm's most eye-catching trilobites. Many of the recovered specimens of this genus—examples of which are also found in both New York State and the English Midlands—display some degree of damage, but their anatomical complexity and aesthetic splendor clearly outshine any superficial structural issues.

HOPLOLICHOIDES FURFICER

In the late 1990s, this was among the first of the "exotic" Ordovician trilobites to invade the international market from the faunally abundant fossil strata of western Russia. Almost immediately these spinose lichids became coveted commodities by collectors and museum officials, both drawn by their incredible morphological intricacy and inherent Paleozoic elegance. Unfortunately, subsequent information revealed that many of these early specimens had been conveniently "doctored" to make their profusion of spines appear even more prominent and dramatic. However, even with a somewhat reduced degree of spinosity, *Hoplolichoides* remains a singularly notable member of the trilobite class.

KONEPRUSIA DAHMANI

It wasn't until late in the twentieth century when the fossil preparator's skill had reached an advanced level of proficiency that the true magnificence, elegance, and complexity of many Moroccan trilobites became fully evident. That is particularly true for the bizarre Devonian species *Koneprusia dahmani*, a trilobite that highlighted an elongated anatomical profile as well as a notably spinose carapace—a form that may have enabled these creatures to gently float amid powerful oceanic currents. Some *Koneprusia* even presented a highly detailed spine-on-spine preservation, making their Paleozoic morphology more bizarre, more evolutionarily confounding, and even more appealing to dedicated collectors across the face of the planet.

OLENOIDES SUPERBUS

These majestic "monsters" were apparently the scourge of the Middle Cambrian oceans. Frequently reaching lengths of 14 centimeters or more, the beautifully preserved specimens that emerged from Utah's famed Marjum Formation may have been predatory in nature, most likely feasting on smaller, less intimidating members of the trilobite class as well as on a wide variety of soft-bodied faunal arthropods. The row of long spines appearing down the species' axial lobe may have been used for defensive purposes, or possibly as a rudder for navigating through the early Paleozoic seas.

PSYCHOPYGE TERMIERORUM

Of all the celebrated trilobites that have emerged from the bountiful Devonian Period outcrops of Morocco in the past four decades, perhaps none is more instantly recognizable than the large (up to 15-centimeter-long) sword-nosed *Psychopyge*. With a genus name that roughly translates as "crazy tail," few bugs can compete with this extraordinary species in renown or elegance. Perhaps more important, with top-quality specimens of this distinctive trilobite readily available on the commercial market, nice examples are easily obtained by most interested collectors—just make sure you don't acquire one that has been heavily restored or overly "enhanced."

GABRIELLUS KIERORUM

It is abundantly apparent that nature's evolutionary forces wasted little time when it came to casting their pervasive influence on the trilobite world. Even during the initial stages of the Lower Cambrian, soon after trilobites first emerged some 521 million years ago, wonderfully detailed and elegantly configured species such as *Gabriellus kierorum* were already filling the seas. The impressive size (up to 12 centimeters in length) and morphological complexity of these specimens provide striking evidence of the trilobite class's rapid development in that primal marine ecosystem.

BOEDASPIS ENSIFER

When these magnificent, spine-covered Russian trilobites first began to appear at international fossil shows in the early years of this century, few in the scientific and collecting communities could believe their arthropod lovin' eyes. Initially it was thought that the large odontopleurid *Boedaspis* (often exceeding 10 centimeters in length) must be the figment of some overzealous prep artist's overactive imagination. Thankfully, as more specimens emerged, the status of these trilobites as true-blue (actually an attractive toffee-toned calcite) Ordovician relics was confirmed—although it was subsequently revealed that more than a few of these specimens had been artificially enhanced with shell-color-matching doses of epoxy and resin.

REMOPLEURIDES ASTEROIDEUS (REED, 1935) AND DENDROCYSTOIDES SCOTICUS (BATHER, 1913)

Upper Ordovician, Ashgill Series, Rawtheyan Stage; Upper Drummuck Group, South Threave Formation; Farden Member, Starfish Bed A; Lady Burn, Girvan, Ayrshire, Scotland; 4.5 cm

This piece was collected back in the 1930s.

***GABRICERAURUS DENTATUS* (RAYMOND AND BARTON, 1913)**

Upper Ordovician; Bobcaygeon Formation; Deseronto, Ontario, Canada; 10.8 cm

(LEFT) *KONEPRUSIA SP.*

Lower Devonian, Emsian Stage; Hmar Lakhdad, Morocco; 2.9 cm

(BOTTOM) *OLENOIDES SUPERBUS* **(WALCOTT, 1908)**

Middle Cambrian; Marjum Formation, House Range; Millard County, Utah, United States; Dorsal (top) 10.6 cm, ventral 12.5 cm

This is a unique dorsal/ventral "double" of this rare species.

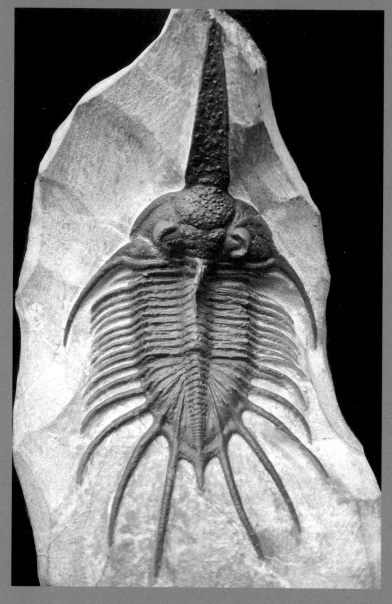

(TOP) *PSYCHOPYGE TERMIERORUM* MORZADEC, 2001

Lower Devonian, Upper Emsian; Tazoulait Formation; Jebel Issoumour, Alnif, Morocco; 12.7 cm with spines

(RIGHT) *GABRIELLUS KIERORUM* (CORBACHO AND LOPEZ-SORIANO, 2013)

Lower Cambrian; Rosella Formation, Atan Group; Dease River, Northern British Columbia, Canada; 11.2 cm

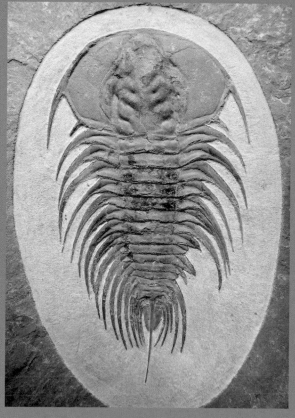

14

10 TIME AND TRILOBITE-RELATED THEORIES

Those who collect and study trilobites have never been particularly intimidated by large numbers or controversial theories. After all, in a field in which tens of thousands of species are scientifically recognized and hundreds of millions of years represent the standard means of time measurement, a well-honed "feel" for mind-numbing numbers, as well as a secure sense of the academically abstract, appears to go right along with the Paleozoic territory. Coming into daily contact with primitive and occasionally hard to pronounce trilobite species such as *Olenoides marjumensis, Conocoryphe sulzuri*, and *Mesonacis georgiensis*—whose half-a-billion-year-old origins date back to the very beginning of complex life on our planet— seems to place trilo-centric folks squarely in their Cambrian comfort zone. Whether it's due to the atypical nature of their peculiar interest or is merely the by-product of a healthy imagination, those fascinated by trilobites appear to possess a decidedly different lens through which to view both the movement of time and the role that these engaging arthropods played in that ever-changing construct. There seems to be little doubt that those who choose to sequester themselves in a realm where trilobites dominate their thoughts, deeds, and actions are willing to stare the great abyss of time squarely in its fanciful face—and then live to tell about it. As they deal with an apparently never-ending supply of cerebrally challenging topics, trilobite enthusiasts also invariably come in direct contact with a variety of scientifically scintillating theories that touch on time's impact on the world's favorite fossil arthropod.

AMPHILICHAS SP.
Upper Ordovician, Ashgillian; Ddolhr Beds; Cynwyd Forest,
Wales, United Kingdom; 4.1 cm

With all of that in mind, here are 10 of those time and trilobite-related theories.

PLATE TECTONICS

This widely hailed concept (previously recognized in some scientific circles as continental drift) conveniently explains why you can find the remains of remarkably similar trilobite species in locations now thousands of kilometers apart. Yes, the answer to any and all such issues can be directly linked to plate tectonics. For hundreds of millions of years, the planet's continental masses have been slowly sliding across Earth's underlying lithosphere like supersized hotcakes on a well-greased skillet. As

they proceed on their millimeter-by-millimeter journey, they carry along the fossilized flora and fauna (certainly including trilobites) contained within their well-stratified sedimentary layers.

PUNCTUATED EQUILIBRIA

Since the mid-nineteenth-century days of Charles Darwin, it had been speculated—and generally accepted—that evolution was a rather slow and steady process. A colony of birds would land on an island, somehow get stranded, and yada, yada. When the paleontologists Niles Eldredge and Stephen J. Gould presented the theory of punctuated equilibria to the world in 1972, they turned many of Darwin's hallowed evolutionary concepts upside down and inside out. Based in significant part on Eldredge's study of phacopid trilobites, they postulated that at its core evolution occurs in a more fits and starts manner. After lengthy periods of relative stasis, a species could rather quickly evolve to meet any new environmental conditions and the demands that were placed on it.

SNOWBALL EARTH

In the later years of the twentieth century, scientists suggested that at some point significantly prior to the dawn of the Cambrian Period—possibly as far back as 700 million years ago—the entire globe had become entrapped in a layer of ice. This phenomenon has since become popularized under the easy to remember moniker "Snowball Earth." Academics have long known that our world has been subjected to fluctuating periods of glaciation and subsequent warming. To the best of their admittedly limited knowledge, however, none of these previous ice ages featured a stage during which the entire planet was covered in a thick coating of pole-to-pole permafrost. The end of this "snowball" phenomenon led indirectly to the famed Cambrian Explosion and to the eventual emergence of trilobites.

TRILOBITE CARAPACES

Over the last few decades, a number of academically inclined theories have been offered to conveniently explain the emergence of the trilobites' characteristic outer armor—its shell. It seems safe to speculate (judging by the fossil record) that all trilobite precursors were soft-bodied organisms and that the development of any subsequent calcite-covered carapace was due to one overriding reason—the evolutionary advantage a hard protective shell provided in an increasingly hostile marine ecosystem. It has also been hypothesized that such an outer casing may have protected certain early, shallow-water trilobite species from the harsh rays of the Paleozoic sun.

OLENOIDES MARJUMENSIS
RESSER, 1942
Middle Cambrian, Marjumian; Marjum Shale; House Range, Millard County, Utah, United States; 6.8 cm

TRILOBITE TRAVEL

In direct contrast to the public perception that trilobites were solitary creatures—a notion perpetuated by the generally singular nature of their fossilized remains—it appears that many trilobite species were highly communal animals. In fact, a surprising degree of fossil evidence reveals that these creatures often congregated in tightly packed groups, perhaps even traversing the world's seafloors in long, single-file, cephalon-to-pygidium alignments. This behavior was apparently designed to provide safety in numbers as well as to subsequently provide a marked increase in each trilobite's reproductive opportunities.

TRILOBITE GROWTH

In recent years, a variety of papers, dissertations, tomes, and treatises have been presented that attempt to explain the incredible size variance that exists within the trilobite class. Of course, it can be surmised that any group of organisms that existed for more than a quarter-billion-year span should be expected to present a widely divergent paleontological profile. With adult trilobite sizes ranging from a few millimeters to more than 70 centimeters in length, scientific speculation has long conveyed the concept that marine water temperature, potential predators, and the ecological environment in which these creatures lived were primary factors that dictated the potential dimensions any trilobite could attain.

TRILO-POOP

What? Yup, you read that correctly—trilo-poop. Indeed, it has been academically hypothesized that as far back as the Lower Cambrian the natural fertilizer derived from the excrement of these omnipresent invertebrates may have been pervasive enough to play a key role in creating marine environments more hospitable to the emergence of primal plant life. This rapid vegetative expansion subsequently created more carbon dioxide in the oceans, which also served to eventually make Earth's atmosphere more amenable for terrestrial life. With a bit of a stretch, it can be speculated that all of today's extant species (including humans) might not be around if it weren't for this distant and obscure bit of trilobite-derived biological reckoning.

MOLTING HABITS

A recent scientific paper proposed that the trilobite class's antiquated molting behavior may have served as the primary impetus behind their slowly decreasing numbers throughout their prolonged Deep Time passage. It has also been academically argued that once trilobites developed the technique for regularly discarding their hard calcite shells—a necessary part of the growth process—they never significantly altered their basic means of molting. However, as their morphological complexity increased over time, such a well-established behavior apparently began to severely limit their adaptive capabilities.

COMPOUND EYES

Ten million years before development of the trilobites' characteristic calcite shells, it appears that certain members of the arthropod line already possessed complex compound eyes—an evolutionary advantage that provided them with a unique view of the undersea world that surrounded them. Such clear-cut optical evidence has yet to be revealed through the fossilized remains of any recognized trilobite precursors, but specimens of other early arthropods displaying this major morphological advantage—including those of the legendary

Anomalocaris—have been found in the Lower Cambrian Emu Bay Shale of South Australia. These poorly preserved specimens graphically show that some arthropods already featured large, multilensed compound eyes even before their emergence in trilobites.

EVOLUTION

For far too many inhabitants of planet Earth, evolution remains the archetypal elephant in the room—a "theory" that nearly half of the American populace still seems unwilling to acknowledge or accept. Yet the indisputable fact that trilobites emerged more than half-a-billion years ago, survived for 270 million years, and produced more than 25,000 known species should be more than sufficient to convince at least *some* of those incredibly stubborn, misguided, or misinformed folks that evolution is *not* some hair-brained concept. Evolution is a reality as fundamental to life on this planet as the air we breathe and the water we drink.

MESONACIS GEORGIENSIS (RESSER AND HOWELL, 1938)

Lower Cambrian; Rome Formation, Montevallo Shale; Montevallo, Alabama, United States; 2.8 cm

***PLATYCALYMENE DUPLICATA* (MURCHISON, 1839)**

Upper Ordovician, Caradoc Series, Aurelucian Stage; Upper Llanfawr Mudstone Formation; Llandrindod Wells, Powys, Wales, United Kingdom; Larger specimen 3.4 cm

(TOP) *RIELASPIS ELEGANTULA* (BILLINGS, 1866)

Lower Silurian; Thornloe Formation; Lake Temiskaming area;
Northern Ontario, Canada; 2.7 cm

**(RIGHT) *MACROPYGE (PROMACROPYGE) CILIENSIS*
PENG, 1984**

Lower Ordovician, Tremadocian; Guotang Formation; Luxi
County, Hunan Province; China; 6.7 cm

Photo courtesy of the Shengpeng Li Collection

15

10 MOST COMMON TRILOBITES

Contrary to popular belief, not every trilobite is a rare and scientifically significant treasure. In truth, the majority of specimens found in both museum exhibits and private collections are little more than relatively common (and hopefully complete) remnants of the distant Paleozoic past. With a minimum of effort—and a surprisingly modest budget—anyone with even the slightest degree of interest should be able to find inexpensive examples of these often abundant trilo-types from Morocco, China, Europe, and North America lurking at local rock shows, displayed in museum gift shops, or making their presence known all over the internet. In recent years—thanks, in part, to feature articles on this fossiliferous subject appearing in such prestigious outlets as *Forbes* magazine and the *New York Times*—trilobite collecting has developed a somewhat undeserved reputation as being a rich man's pursuit. By acquiring the globe-spanning species featured here, even the most prudent armchair collector can quickly, easily, and rather inexpensively (with most specimens costing under $100) gather together a noteworthy assemblage of these ancient arthropods—one guaranteed to serve as a major source of enjoyment and as a point of conversation among visiting friends and family.

Here is a brief salute to the 10 most common trilobites in the world.

ILLAENULA VIETNAMICA

When these midsized (3- to 4-centimeter-long) Middle Devonian trilobites began invading the natural history scene late in the twentieth century, they were originally marketed as *Ductina vietnamica*. Unfortunately, neither that genus nor the notion that they hailed from Vietnam proved to be correct. In fact, the elliptically shaped *Illaenula* emerged from a dig site in the Guangxi Province of southern China, some 900 kilometers from the Vietnamese border. Hundreds, if not thousands, of these distinctive arthropods have been unearthed since their initial discovery, resulting in their pervasive appearance at fossil shows as well as on eBay.

ILLAENULA VIETNAMICA (MAKSIMOVA, 1965)
Middle Devonian, Eifelian; Nabiao Formation; Hunan Province, China; largest trilobite 4 cm

ELRATHIA KINGII

This ubiquitous Middle Cambrian species dominates the fossil fauna in the 510-million-year-old Wheeler Shale of Utah. It has been estimated that more than 100,000 *Elrathia* specimens have been found, making this attractive, black, brown, or red-shelled 2- to 4-centimeter-long species the world's most common trilobite. But be aware that most available examples lack their free cheeks; it pays to take an extra moment and fork over a few extra coins to procure a relatively pristine specimen.

ELLIPSOCEPHALUS HOFFI

Often found in mass mortality layers featuring scores of these diminutive (2-centimeter) trilobites overlapping one another in a veritable Paleozoic heap, the Middle Cambrian representative *Ellipsocephalus* is unquestionably the most prolific species emerging from the famous Barrandian Lagerstatte of the Czech Republic. Indeed, this is perhaps the most common trilobite species in all of Europe. It also is among the first scientifically recognized trilobites and was named in 1823.

FLEXICALYMENE OUZREGUI

These large (up to 10-centimeter-long) Ordovician "mud bugs" come in easily breakable concretions that traditionally feature both positive and negative halves of the trilobite. This species of *Flexicalymene* has been pervasive since the beginning of the Moroccan trilobite boom in the late 1980s. Frequently found at rock and mineral shows, as well as across the internet, be aware that many "complete" examples have been fabricated from composited, disjointed, unassociated pieces, making them an ultra-common pariah in the commercial marketplace.

ASAPHUS EXPANSUS

With their beautifully preserved toffee-toned shells, the medium-sized (8- to 12-centimeter) *Asaphus expansus* may be the most readily available example of the often lavish line of trilobites drawn from western Russia's bulging Ordovician fossil beds. However, be aware that various levels of restoration may characterize even the most common species emerging from the quarries that ring St. Petersburg. In recent days, these practices seem to have become more limited to top of the line, exotic (and expensive) trilobite specimens.

ELDREDGEOPS RANA

This 2- to 3-centimeter-long species from the Devonian rocks of upstate New York (formerly known as *Phacops rana*, which, for those wondering, roughly translates as "frog face") can sometimes appear in mass mortality plates featuring up to 100 tightly packed complete examples. In fact, this is the most common trilobite found throughout the northeastern United States. With their coal-black shells and amazingly detailed compound eyes, this classic trilobite ranks high among collectors' favorites.

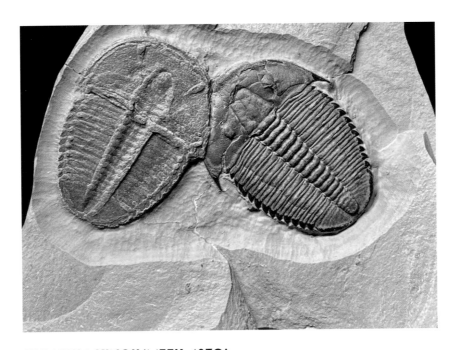

ELRATHIA KINGII (MEEK, 1870)
Late Middle Cambrian; Wheeler Formation—base of Marjum Formation; House Range, Utah, United States; Each trilobite 2.3 cm; Note antennae inside the cephalon of left specimen.

Photo courtesy of Markus Martin

ASAPHISCUS WHEELERI

Asaphiscus wheeleri represents another of Utah's classic and omnipresent Middle Cambrian species. Although not as common as *Elrathia kingii* (nothing else in the trilobite world is), this is still the second most abundant nonagnostid genus known from the Wheeler Shale, and during the last century thousands of *Asaphiscus* have been recovered. Complete examples of this 4- to 7-centimeter species are much more difficult to obtain than molted specimens lacking their free cheeks. These trilobites often can be found in attractive colors ranging from red to brown to black.

CHANGASPIS ELONGOTA

This diminutive but elegant Lower Cambrian species from China started popping up on the global fossil market in significant numbers during the early days of the twenty-first century. In recent years, good examples of the surprisingly spinose, 2-centimeter-long *Changaspis* have become more and more difficult to find through mainstream commercial outlets. These trilobites still appear on eBay and other internet sites and can generally be obtained for $20 or less.

FLEXICALYMENE MEEKI

To state something somewhat oxymoronic, *Flexicalymene meeki* is probably the "rarest" member on this "most common" trilobites list. Examples of this popular Ordovician species from Ohio—often discovered in an enrolled state of preservation—are still being found in sufficient numbers to make them essential and expected components for just about every private or museum trilobite collection. Indeed, numerous midwestern hobbyists specialize exclusively in these well-preserved Paleozoic remnants, and many of them proceed to fill everything from cabinet shelves to mason jars with their numerous discoveries.

GERASTOS GRANULOSUS

These small, ovate trilobites rank as perhaps the most common of all Devonian-age Moroccan species. *Gerastos* can usually be found being marketed by the dozens at virtually every major fossil show across the globe. One must assume that the cost of Moroccan prep is significantly lower than its Euro-American equivalent, otherwise these "cute," three-dimensional, 2-centimeter-long proetid specimens would never be available for their expected price tag of only a few dollars each.

(TOP, LEFT) *ELLIPSOCEPHALUS HOFFI* **(SCHLOTHEIM, 1823)**

Middle Cambrian; Jince Formation; Litavka River Valley, Czech Republic; largest trilobite 2 cm

(BELOW, LEFT) *ASAPHUS EXPANSUS* **(WAHLENBERG, 1821)**

Middle Ordovician, Upper Arenigian; Kunda Regional Stage; Lynna, Sillaoru Formations; Voybokalo, St. Petersburg region, Russia; 8.6 cm

(BOTTOM) *ASAPHISCUS WHEELERI* (MEEK, 1873)

Middle Cambrian; Wheeler Formation; Utah, United States; 6.5 cm

***ELDREDGEOPS RANA* (GREEN, 1832)**

Middle Devonian, Givetian; Hamilton Group, Moscow Formation, Windom Shale Member; Penn Dixie Quarry; Hamburg, New York, United States; average size 3 cm

16

10 RIDICULOUSLY RARE TRILOBITES

Many trilobites are pervasive, having been discovered by the scores in various Paleozoic layers of planet Earth. Others are as rare as precious gems—or rarer! In some cases, fossilized examples of these ancient arthropods represent the *only* complete specimen ever found of that particular species or genus. Ironically, it is often these unique representatives of the distant past that best demonstrate the extraordinary diversity inherent in the trilobite class. The incredible specimens presented in this section—including *Platylichas laxatus*, *Dikelocephalus minnesotensis*, and *Acanthopyge n. sp.*—haven't necessarily been chosen to reflect some highbrow paleontological principle, highlight the contents of a hallowed quarry, or reveal some previously unrecognized anatomical feature. Rather, they're being discussed and displayed merely because of each trilobite's stunning strangeness, inherent beauty, and remarkable rarity—qualities that make each one more than worthy of being seen, studied, collected, and admired. In recent years, more and more of these distinctive specimens have been revealed lurking amid properly aged sedimentary outcrops around the globe, whether that locality is a desert-adjacent escarpment in North Africa or a snow-capped mountain range in Bolivia. With each passing day, as explorers dig deeper into the nooks and crannies found in the planet's most remote corners, additional examples of these previously unknown and unseen trilobite species continue to be unveiled to the amazement and pleasure of enthusiasts everywhere.

PLATYLICHAS LAXATUS (McCOY, 1846)
Upper Ordovician, Caradoc Series; Streffordian Stage, Actonian Substage; Acton Scott Formation; Cardington, Shropshire, England; 6.5 cm

In recognition of many of those singular discoveries, here are 10 of the rarest trilobites in the world.

PLATYLICHAS LAXATUS

The rolling green hills of Shropshire in the English Midlands have long provided a significant (although not always easy to either locate or extricate) storehouse of both Ordovician and Silurian trilobite treasures. Among the area's Ordovician-age outcrops, rare, often only partially preserved examples of exotic species such as *Platylichas laxatus* have been discovered, with these usually appearing in a heavy, beige-tinged, limestone matrix. The few complete examples that have been extracted can require more than 20 hours of careful preparation to be cleared from their surrounding stone casing.

TERATASPIS GRANDIS

This trilobite is the legendary Devonian marine "monster" of eastern North America. Originally described only from disarticulated fragments, *Terataspis* represents a lichid genus that could have grown to dimensions of 45 centimeters or more. One complete Canadian example is currently on display in the Royal Ontario Museum; another rumored articulated specimen was found in northern New York State early in the twenty-first century and is currently the feature attraction of a noted Japanese collection.

LAETHOPRUSIA SP.

Oklahoma's Haragan Formation has produced some of the best known—and best preserved—Devonian trilobites ever found. But even after decades of extensive digging at the famed Black Cat Mountain locality, surprises can still occur. Such was the case when a single example of a previously unknown species of *Laethoprusia* was discovered there in 2007. This 2-centimeter-long odontopleurid is noted for the delicate, sweeping spines that encircle the trilobite's carapace. A similar—and almost as rare—example of the genus (although some academics believe it may be a specimen of the closely related *Isoprusia*) is recognized from similarly aged strata in Morocco.

DICRANURUS BARBURUS

A handful of these striking—although still not officially identified—trilobites have been uncovered in the centuries since English explorers first

LAETHOPRUSIA SP.

Middle Devonian, Eifelian; Bou Tchrafine Formation; Jorf, Morocco; 3 cm

began active explorations of the Silurian-age Much Wenlock exposures of Shropshire. Yet it has only been in the early years of the current millennium that complete examples of what is being called *Dicranurus barburus* (or *Selenepeltoides barburus*, depending on your source material) have become available on the global stage. Most of these well-preserved specimens have been less than 5 centimeters in length, but disarticulated fragments indicate that these trilobites could have grown considerably larger.

BREVIREDLICHIA GRANULOSA

Judging by the evidence provided by bits and pieces of fossilized trilobite anatomy uncovered in China's 515-million-year-old Wulongqing Formation, examples of the distinctive Lower Cambrian trilobite *Breviredlichia granulosa* may have grown to 20 centimeters in length. The largest complete specimens yet found have measured 12 centimeters from cephalon to pygidium, and those scarce trilobites are almost as wide due to their dramatically extended genal spines. The species name is drawn from the granulose texture that covers much of the carapace of these impressive and rare arthropods.

METOPOLICHAS BREVICEPS

Found in the famed Silurian-age Waldron Shale quarries that mark the Indiana landscape, only one complete 6-centimeter-long example of the elegant lichid *Metopolichas breviceps* has been reported, although tantalizing fragmented remains are relatively common finds throughout the vicinity. Apparently, strong oceanic currents prevailed in this locale some 425 million years ago, serving to tear apart many of the area's trilobites (as well as their frequently molted exoskeletons) soon after their demise.

PARANEPHROLENELLUS BESTI

An often told story in certain collecting circles describes a unique 4-centimeter specimen of *Paranephrolenellus besti* that was acquired in the late-1990s from an American vendor who was roaming aimlessly through the Tucson Fossil Show while transporting the fragile trilobite in his backpack. Exactly where *he* acquired it remains a bit of a Paleozoic mystery, although some believe it may have been self-collected decades earlier. That trilobite represents a distinctly colorful and notably rare example of a species drawn from the Lower Cambrian Carrara Formation in southeastern California.

DIKELOCEPHALUS MINNESOTENSIS

For more than two decades—until the revamping of its fossil halls in 2016—this specimen of *Dikelocephalus minnesotensis* served as the centerpiece of the Smithsonian's renowned trilobite display. Supposedly "discovered" in the 1960s while it was being used as a doorstop in a midwestern home, that 8-centimeter-long specimen from the Upper Cambrian of Wisconsin represents one of the only known intact examples of this distinctive species, which had initially been recognized and scientifically described from their commonly found disarticulated pygidia.

OLENOIDES SKABELUNDI

Dating back to the initial fossil material found in this picturesque western site more than a hundred years ago, it was assumed that the exceptional trilobite fauna hailing from Utah's Weeks Formation was Upper Cambrian in age. However, during the early years of the current century, two virtually complete *Olenoides skabelundi* specimens (one nearly 18 centimeters in length) were revealed.

BREVIREDLICHIA N. SP.
Lower Cambrian; Balang Formation; Hunan Province, China; 4.7 cm

These discoveries provided ample evidence that the Weeks layers were, in fact, among the last remnants of the Middle Cambrian to be found on the North American continent.

URALICHAS AFF. GUITIERREZI

In recent years, a key outcrop of the Valongo Formation that emerges near the town of Arouca, Portugal, has become renowned for the impressive sizes attained by the trilobites found in its smooth, charcoal-toned slate layers. Perhaps the rarest example emanating from this Middle Ordovician site known to locals as the Canelas Quarry is an exceptional species of *Uralichas*, a genus also recognized from Spain and Morocco. Judging by both the three known articulated Portuguese specimens and a variety of disarticulated fragments, this species could have grown to more than 35 centimeters in length.

***DICRANURUS BARBURUS?* THOMAS, 1981; "*SELENOPELTOIDES*" *BARBURUS* (THOMAS, UNPUBLISHED); WITH *DALMANITES CF. IMBRICATULUS* (ANGELIN, 1851)**

Middle Silurian, Wenlockian; Coalbrookdale Formation, Lower Farley Member; Much Wenlock, Shropshire, England; Dicranurus 3.7 cm

This species is still not officially identified, thus the "?"

(TOP, LEFT) *PARANEPHROLENELLUS BESTI*
WEBSTER, 2007

Lower Cambrian; Carrara Formation, Echo Shale
Member; Emigrant Pass, California, United States;
3.7 cm

(TOP) *DIKELOCEPHALUS*
MINNESOTENSIS (OWEN, 1852)

Upper Cambrian; St. Lawrence Formation; Sauk
County, Wisconsin, United States; 12.3 cm

(BELOW, LEFT) *ACANTHOPYGE N. SP.*

Middle Devonian, Eifelian Stage; Bou Tchrafine
Formation; Jorf, Morocco; 4 cm

***CONOMICMACCA ALTA* (LIÑAN & GOZALO, 1986)**

Middle Cambrian, Tissafinian Stage; Jebel Wawrmast Formation, Breche a Micmacca Member; Jebel Ougnate, Tarhoucht, Morocco; 16.5 cm

Photo courtesy of the Sam Stubbs Collection

17

10 BEAUTIFUL BOHEMIAN TRILOBITES

Initial research on the trilobites of Bohemia (now the Czech Republic) was conducted in the mid-nineteenth century by one of paleontology's towering figures, Joachim Barrande. This French-born naturalist began his fieldwork in a broad expanse of 510-million-year-old outcrops that have become known as the Jince Formation. Barrande found this Middle Cambrian exposure to be geologically dynamic. A peculiar layer-cake stratification—with bands of fossil-rich shale interspersed between generally barren sedimentary planes—reflected the unpredictable climatic conditions that affected the region during this period of Earth history. These special environmental circumstances nurtured the emergence of an incredibly diverse trilobite fauna, and more than 60 species of these ancient arthropods have been found in the Jince's bountiful strata. The most common of these include *Paradoxides gracilis*, *Ellipsocephalus hoffi*, *Conocoryphe sulzuri*, and *Hydrocephalus minor*; the small, 2-centimeter-long *Ellipsocephalus* have been recovered by the thousands. The region's trilobites are most often found in a heavily compacted, chocolate-colored mudstone. When pried out of the area's fossiliferous layers and broken apart by chisel and hammer, the larger Jince Formation trilobites (which include *Paradoxides* and *Hydrocephalus* species up to 20 centimeters in length) traditionally emerge in rather flat, positive/negative splits. The well-preserved but at times disarticulated exoskeletons of these impressive examples display a thin calcite shell whose color frequently blends with the cocoa-tinged tone of the surrounding matrix.

CHEIRURUS INSIGNIS BEYRICH, 1845

Middle Silurian; Wenlock Series; Liten Formation, Motol Member; Loděnice (Beroun District), Czech Republic; 7.2 cm

Complete specimens have never been particularly prolific at any of the Jince sites, but nearly two centuries of digging in the vicinity's outcrops has produced perhaps the most comprehensive and detailed trilobite assemblage found anywhere in the world.

Here is a look at 10 incredible Bohemian trilobites.

CHEIRURUS INSIGNIS

It is interesting to note that the depth and scope of Joachim Barrade's efforts with Bohemian trilobites, such as the Silurian species *Cheirurus insignis*, apparently had a significant influence on Charles Darwin, who referenced Barrande's pioneering work in his own historic writings. Throughout the later years of his life, Barrande was a vocal proponent of Georges Cuvier's then-popular "catastrophe theory" of change, which postulated that short, violent events—such as earthquakes and volcanic activity—altered the face as well as the fauna of the planet. It is ironic that this concept ran directly against Darwin's own evolutionary beliefs.

EOHARPES BENIGNENSIS

The world's awareness of Bohemian trilobites dates back more than 250 years. The first published reports on the assorted fossils (including a trilobite pygidium) that had been discovered throughout the Prague region appeared in 1770. Various subsequent scientific manuscripts made their appearance during the nineteenth century, with these either directly or indirectly referencing the area's vast paleontological reserves—including exotic Ordovician trilobite genera such as *Eoharpes benignensis*. These printed efforts were, of course, highlighted by Barrande's multivolume work, which is rumored to have personally cost him nearly $15,000 (the equivalent of over $400,000 today) to fashion, publish, and then distribute throughout Europe.

SPINISCUTELLUM UMBELIFERUM

In many world-renowned fossil formations, the fauna and flora provide clues that the surrounding Paleozoic environment was once a warm tropical sea. In contrast, half-a-billion years ago the waters that covered what is now the Lochkov Formation of the central Czech Republic were much cooler. This formation was apparently located significantly south of the equator, and the environment nurtured a wide variety of trilobite species, including *Spiniscutellum umbeliferum*. Such a notably cool marine climate is also evident in other trilobite-laden strata around the planet, including analogous Devonian sites now found in Morocco and eastern North America. To the eyes and ears of leading academics, such information indicates that at this distant time in the planet's history these now-divergent sites may have shared a similar swath of primal ocean floor.

CTENOCEPHALUS CORONATUS

When paleontologists compare the amazing Bohemian trilobite array assembled by Joachim Barrande—which includes *Ctenocephalus coronatus*—to strikingly similar Middle Cambrian trilobite collections discovered in eastern Canada, Sweden, Wales, and Morocco, they provide a foundational building block for one of the most important geological theories of the last century—plate tectonics. That concept describes how the seven major continental plates that comprise our planet's outer shell slowly shifted their lithospheric position over time, sliding across Earth's underlying mantle like frying eggs on a greased skillet.

EOHARPES BENIGNENSIS (BARRANDE, 1872)
Middle Ordovician, Llandeilian; Dobrotivá Formation; Svatá
Dobrotivá, Czech Republic; 2.5 cm

of major European museums, and then often are not complete specimens. Although this 8-centimeter-long trilobite has been subjected to some natural external defacement (lingering elements of its shell can be seen along its fan-shaped pygidium), it still represents a Paleozoic prize of the highest order.

CONOCORYPHE SULZURI

Like many of the Czech Republic's Middle Cambrian trilobites, *Conocorype sulzuri* was one of the species first presented to the public in the richly illustrated treatise *Système Silurien du Centre de la Bohême*. That compendium of Joachim Barrande's lifelong work first began emerging in 1852, and it helped revolutionize the role paleontology played in scientific studies. In fact, the magnificently detailed trilobite drawings presented in the initial volume of that 23-book series—the preponderance of which were created by talented artists and lithographers functioning directly under Barrande's supervision—helped open the eyes of many to the wonders of the Paleozoic world.

BOHEMOLICHAS INCOLA

No matter where they are found—on the snow-swept Siberian plains, the rocky outcrops of Indiana, or the rolling hills of Scotland—members of the Lichida order rank among the most collectible and desired of all trilobites. Some of the rarest of these distinctive specimens hail from the Ordovician and Silurian repositories of the Czech Republic. Indeed, examples of the Ordovician-age *Bohemolichas incola* are usually not seen outside

HYDROCEPHALUS MINOR

The Middle Cambrian trilobites found throughout the Czech Republic represented by *Hydrocephalus minor* are most often revealed in a dense mudstone that runs in color from mid-tan to dark brown. Although these host rocks rarely break cleanly, a stiff hammer blow from a skilled fieldworker can expose just enough of the trilobite to enable additional preparation work to be conducted. By the way, the genus name *Hydrocephalus*, which translates as "water head," is indicative of the large, circular glabella that extends beyond the trilobites' cephalic rim and serves as the most readily recognized morphological feature of this unusual species.

SPINISCUTELLUM UMBELIFERUM (BEYRICH, 1845)

Lower Devonian; Lochkov Formation; Prague, Czech Republic; 7.2 cm

REJKOCEPHALUS ROTUNDATUS

Joachim Barrande offset a great deal of the exorbitant expense involved with writing and publishing *Système Silurien du Centre de la Bohême* by selling many of his best study specimens—some being the trilobites figured in that volume's pages. Most of these examples, including a number of the unusual Middle Cambrian species *Rejkocephalus rotundatus*, eventually formed the core trilobite collection housed in Prague's National Museum. Other original Barrande trilobites

quickly made their way to institutions around the globe. These trilo-centric transactions were headlined by a significant mid-nineteenth-century acquisition by Louis Agassiz for Harvard's Museum of Comparative Zoology, an agglomeration that still represents the most comprehensive collection of Bohemian trilobites housed outside of Europe.

CERATOCEPHALA LOCHKOVIANA

Nearly two centuries after Barrande's historic volumes first made the entire academic community aware of the incredible faunal diversity contained in Bohemia's fossil outcrops, today there is an ongoing renaissance surrounding both his writings and the trilobites that inspired them. These include the rare Devonian species *Ceratocephala lochkoviana*. To celebrate the two-hundredth anniversary of Barrande's birth, in 1999 a limited edition, hard-cover reprint of *Système Silurien du Centre de la Bohême* (volume 1) was issued by the National Museum in Prague, and it almost immediately sold out. That same year, the Czech Republic released a highly popular block of four commemorative stamps, one featuring Barrande's portrait, the other three presenting some of his more unusual trilobite discoveries.

ACADOPARADOXIDES SIROKYI

Barrande's well-publicized marketing of his core trilobite collection—as well as the resulting museum displays featuring unusual Middle Cambrian species such as *Acadoparadoxides sirokyi*—soon triggered a worldwide fascination with these ancient arthropods. Interest escalated, and only a few years after Barrande had finished conducting his pioneering research in the Jince Formation, teams of minimally trained quarrymen were working in many of the same fossil-rich central European exposures. These laborers were renowned (if not reviled) for their willingness to find, piece together, and then sell what were essentially disassociated trilobite parts. In the process of doing so, these financially motivated operatives often created chimera-like monstrosities unknown to science or to the Paleozoic seas.

BOHEMOLICHAS INCOLA (BARRANDE, 1872)

Middle Ordovician; Sarka Formation; Osek (near Rokycany), Czech Republic; 8.3 cm

(TOP) *CONOCORYPHE SULZURI* **(SCHLOTHEIM, 1823)**

Middle Cambrian; Jince Formation; Central Bohemia, Czech Republic; 8.2 cm

(TOP, RIGHT) *HYDROCEPHALUS MINOR MINOR* **(BOECK, 1827)**

Middle Cambrian, Drumian; Jince Formation, *Ellipsocephalus + Rejkocephalus* zone; Jince locality, Czech Republic; 8.7 cm

(BELOW, RIGHT) *CERATOCEPHALA LOCHKOVIANA* **CHLUPAC, 1971**

Lower Devonian; Lochkov Formation; Kotyz u Tmane, Czech Republic; 4.1 cm

18

10 OUTRAGEOUSLY OLD TRILOBITES

The search for the fossilized remains of the first trilobites to swim in Earth's blue waters stands as a primary quest for paleontologists around the globe. Each year new specimens— including *Lochmanolenellus pentagonalis* and *Caspimexis radiatus*—are being uncovered somewhere on the face of the planet, and many of these elementary examples are seemingly destined to push the dawn of the Age of Trilobites even further back into the dusty pages of primal history. However, despite both the continued interest in the subject and their incredibly ancient placement in the fossil record— with some Lower Cambrian trilobites now dating from slightly more than 521 million years ago—finding the first of these Paleozoic relics is far from an easy task. Indeed, it often seems that the older the trilobite the more remote its current location atop the global lithosphere. From the burning deserts of North Africa to the permafrost-covered outcrops of northern Greenland, evidence is currently being sought, and found, about that moment in evolutionary time when trilobites first emerged as the most advanced life-forms Earth had ever produced. Despite the significant role they played in the early history of our world, the planet's sedimentary stratum has guarded the secrets of the oldest trilobites with a steadfast determination. Only a few locations across the globe have so far revealed themselves to be the presumed cradles for these original experiments with complex animal life. In recent years, discoveries made in trilobite-bearing Lower Cambrian outposts such as the Montenegro Formation of Nevada, the Pestrotsvet Formation

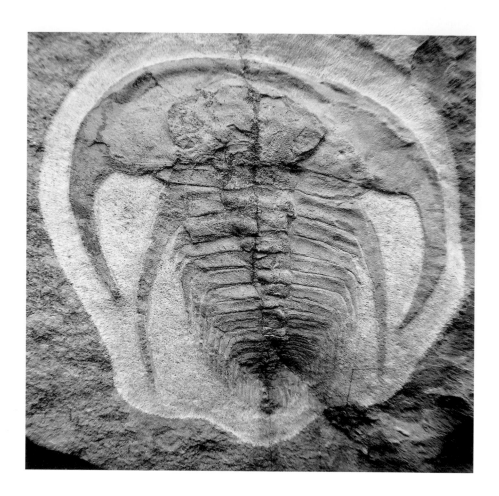

CAMBROINYOELLA CF. WALLACEI LIEBERMAN, 2001
Lower Cambrian; Campito Formation, Montenegro Member; Nevada, United States; 4 cm

of Siberia, and the Chengjiang Biota of China have cast much-needed fossiliferous light on this imperfectly understood period of Earth history. Despite these findings, however, the hunt for the root-member of the trilobite family tree rages on.

The following sections feature 10 of the leading contenders for the prestigious title of World's Oldest Trilobite.

CAMBROINYOELLA WALLACEI

One of the prime Paleozoic contenders for yielding the first member of the trilobite lineage—among an impressive and ever-growing series of candidates—is the 520-million-year-old Campito Formation of the western United States. Rare Lower Cambrian species such as *Cambroinyoella*

wallacei appear to be perched at the very base of the entire, expansive trilobite line. Although rarely found complete, even partial examples dramatically demonstrate the morphological complexity of the trilobite class so early in its march through evolutionary time.

EOFALLOTASPIS TIOUTENSIS

Academic thought concerning the possible birthplace of the trilobite lineage continues to undergo constant updates and revisions. One of the latest candidates for producing these earliest arthropods has emerged amid the Lower Cambrian layers of Morocco, home to many fallotaspid species, including *Eofallotaspis tioutensis*. As fresh fossil material is discovered in that nation's Zagora

Region and new analysis is conducted, paleontologists have slowly begun to garner a more detailed understanding of the circumstances surrounding the emergence of trilobites in Earth's early seas.

LOCHMANOLENELLUS PENTAGONALIS

A key wellspring for early trilobite development may have been located on the ocean shelf adjacent to the ancient microcontinent of Avalonia. In the earliest days of the Cambrian, that landmass rose as a volcanic arc along the northernmost border of the supercontinent called Gondwana. There, around 520 million years ago, on some hospitable shallow-water plateau situated along the equatorial line, a recognized trilobite genus such as *Lochmanolenellus*—whose fossilized remains are now found in the Esmeralda Basin of Nevada—may have reared its antennae-adorned cephalon for the very first time.

BIGOTINA BIVALLATA

Recent scientific studies suggest that the calcite-clad trilobite class may have originally arisen in the waters that once enveloped Siberian exposures in which the fossilized remains of primitive Lower Cambrian species such as *Bigotina bivallata* have been uncovered. Over a relatively short geologic period—perhaps less than 200,000 years—hundreds of divergent trilobite species may have radiated from that single Siberian locale to reach suitable marine environments across the expanse of our world.

ARCHAEASPIS MACROPLEURON

At almost the same time that 521-million-year-old species such as *Archaeaspis macropleuron* were first leaving their mineralized impressions in sedimentary layers now found in Nevada, other early representatives of the trilobite line (including such genera as *Eofallotaspis* and *Fritzaspis*) were emerging in key biological hotspots around the globe. Lower Cambrian outcrops recently uncovered in Spain, Greenland, Morocco, and Siberia may eventually prove to contain the earliest known representative of the noble trilobite class.

DELGADELLA LENAICA

Siberia's Sinsk Formation contains a significant and diverse Lower Cambrian fauna. The nearby and slightly older Pestrotsvet Formation has continually been among the least productive of

LOCHMANOLENELLUS PENTAGONALIS WEBSTER AND BOHACH, 2014
Lower Cambrian, Series 2, Lower Dyeran; Poleta Formation; Esmeralda County, Nevada, United States; 2.8 cm

ARCHAEASPIS
MACROPLEURON
LIEBERMAN, 2002

Lower Cambrian; Campito Formation,
Montenegro Member; Nevada, United
States; 3.1 cm

trilobite repositories, having yielded only a few complete, articulated specimens. But the material in the Sinsk, although certainly not what anyone would label as bountiful, is clearly more abundantly distributed. Trilobites such as *Delgadella lenaica* even occasionally appear as multiple, if imperfectly preserved examples upon a single sedimentary stone. Over the last few decades, fossil enthusiasts around the globe have eagerly sought these prime Deep Time specimens to add to their collections.

JAKUTUS PRIMIGENIUS

In a variety of Siberian locations, the powerful currents of the Lena and Anabar rivers have exposed wide swaths of Lower Cambrian Sinsk Formation strata. Complete—and often impressively large—examples of *Jakutus primigenius*, some up to 15 centimeters in length, have been found there. Almost always preserved with a thin, coffee-hued calcite coating, which beautifully captures the subtle morphological variances displayed by this species, these distinctive fossils have quickly become prime targets of acquisition for trilobite enthusiasts from Moscow to Miami—as well as the subject of growing academic analysis. A substantial number of these 520-million-year-old trilobites display pronounced axial spines, perhaps among the first times such an advanced anatomical feature is evident in the fossil record.

PERRECTOR (RICHTEROPS) FALLOTI

Throughout the last half-century, the diversity of trilobite material that has emerged from North Africa has proven to be nothing less than astounding. From Lower Cambrian to Upper Devonian, thousands of distinct trilobite species (many yet to be scientifically described) have been brought forth from Morocco's rich Paleozoic strata. Among the earliest of these often beautifully preserved arthropods are members of the genus *Perrector* (until recently known as *Richterops*), which have been recovered from long-established quarries located along the nation's eastern coast.

CASPIMEXIS RADIATUS

This beautiful and rarely seen trilobite species hails from a well-known but infrequently explored Lower Cambrian exposure located in the Nevada desert. Such discoveries reveal just how fast the trilobite line was evolving in the primal seas; more than half-a-billion years ago the first trilobites emerged, and a mere instant later in geological time these extraordinary creatures were primed to begin one of the most fertile periods in their entire history. Thousands of new species, including *Caspimexis radiatus* (known previously as *Mexicaspis*), would soon pervade the planet's Cambrian oceans and help firmly establish a class of animals that would survive for the next 270 million years.

SERRANIA SP.

Despite extensive and ongoing academic explorations, no complete examples of the diminutive genus *Serrania* have yet been unearthed, although disarticulated fragments of this early trilobite have been found in various outcrops located across the Iberian Peninsula. These discoveries indicate that the Lower Cambrian seas were virtually bursting with evolutionary energy, especially when we consider the breadth and scope of the trilobites hailing from the 521-million-year-old sedimentary layers of Spain, China, Morocco, and Siberia. Perhaps at no other time in our planet's 4.54-billion-year history have the forces of nature been more inventive or unpredictable.

JAKUTUS PRIMIGENIUS IVANTSOV, 2005

Lower Cambrian; Achchagy-Tuoydakh Lagerstatte; Southern Yakutia, Siberia; 1.2 cm

(TOP, LEFT) *ELLIPTOCEPHALA SP.*

Lower Cambrian; Poleta Formation; Esmeralda County, Nevada, United States; 6.7 cm

(TOP, RIGHT) *CASPIMEXIS RADIATUS* PALMER AND HALLEY, 1979

Lower Cambrian; Carrara Formation; Pahrump, Nevada, United States; 3.3 cm

(BELOW, LEFT) *PERRECTOR (RICHTEROPS) FALLOTI* HUPÉ, 1953

Lower Cambrian, Series 2, Age 3, Issendalenian Regional Stage; Amouslek Formation, Daguinaspis and Resserops Zones; Taz Emmour, Morocco; 4.7 cm

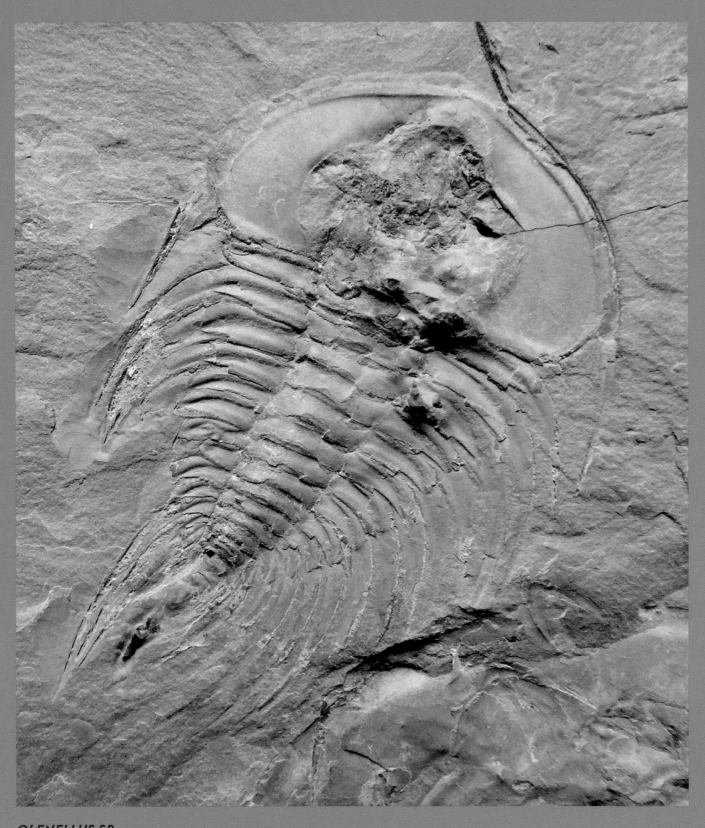

OLENELLUS SP.

Lower Cambrian, Series 2; Atan Group, Rosella Formation; Dease River, British Columbia, Canada; 7.2 cm

19

10 LAST-IN-LINE TRILOBITES

T rilobites crawled through Earth's early oceans for more than a quarter of a billion years. Following a slow yet steady decline that began not long after their emergence in the Cambrian, by the end of the Permian Period the trilobite class had reached the conclusion of its lengthy passage through primal history. A variety of environmental factors have been presented as the root cause behind their precipitous downturn: changing marine habitats, more menacing predators, celestial gamma ray bursts, undersea methane eruptions, or even problems caused by their increasingly outdated molting behavior. There is ample documentation in the fossil record showing that the entire trilobite line was in dire evolutionary straits well before its eventual demise. Determining which naturally occurring events presented the primary impetus that finally pushed these incredibly adaptable arthropods beyond the brink of recovery remains something of a Paleozoic mystery. One recent theory postulates that a string of supersized volcanoes simultaneously erupted throughout the region that is now Siberia 252 million years ago, filling both air and sea with enough carbon dioxide residue to slowly suffocate terrestrial and marine lifeforms everywhere on Earth. Other academics believe that an asteroid or meteorite shower—possibly consisting of up to a dozen large-scale, iron-nickel alloy space rocks, which together would have unleashed the power of over a million nuclear warheads—may have simultaneously struck Earth at this critical juncture in the planet's history. Whatever the cause, the aftermath was uniformly devastating for life around the globe, resulting in the largest mass extinction in our world's 4.54-billion-year history. More than 96 percent of marine species—including *all* remaining trilobites, which by this time were exclusively of the Proetida order—and 70 percent of terrestrial life-forms perished in that event's prolonged wake.

In careful consideration of that planet-changing, trilobite-ending episode, here's a look at 10 last-in-line trilobites.

AMEURA MAJOR

These rare, midsized (3- to 5-centimeter-long) trilobites are uncovered in 320-million-year-old

AMEURA MAJOR (SHUMARD, 1858)
Late Pennsylvanian (Missouri age); Shawnee Formation; Doniphan Shale bed of Lecompton Limestone Member; Missouri, United States; 3.2 cm

Pennsylvanian-age pockets throughout the American Midwest, with the best-known examples emanating from formations in Nebraska, Missouri, Illinois, and Kansas. Due to their attractive hydrodynamic shape and lovely beige calcite exoskeletons, *Ameura* rank among the most coveted of all end-of-the-line trilobites. Indeed, collectors from St. Louis to St. Petersburg are perpetually scrambling to add a complete example of this unusual genus to their Paleozoic holdings.

PHILLIPSIA ROCKFORDENSIS

By the dawn of the initial stage of the Carboniferous (the Mississippian Period), the few remaining trilobites—all members of the Proetida order—were both uniformly small (usually less than 5 centimeters in length) and uniformly shaped. The uncommon species *Phillipsia rockfordensis* hails from the Borden Rock Group of Brown County, Indiana. There the recovered examples are often disarticulated and found in mudstone concretions that wash out of the 350-million-year-old strata that aligns with a neighboring riverbed.

NAMUROPYGE NEWMEXICOENSIS

The Lake Valley Formation of southwestern New Mexico was first explored in the later years of the nineteenth century by the famed paleontologist Edward Drinker Cope—a man more historically renowned for his interest in dinosaurs than in trilobites. At that time, his work uncovered only partial examples of distinctive species such as *Namuropyge newmexicoensis*. Subsequent explorations of the area have revealed a veritable cornucopia of complete trilobite types, many of which are beautifully preserved in a thick, dark-brown calcite. In fact, the Lake Valley Formation presents the most diverse post-Devonian trilobite fauna in the world.

PHILLIPSIA ROCKFORDENSIS WINCHELL, 1865
Lower Mississippian; Borden Rock Group; Brown County, Indiana, United States; 6.1 cm

DITOMOPYGE PRODUCTA

Perhaps no place on planet Earth contains more Paleozoic outcrops (ranging from Cambrian to Permian) than that mysterious land known as Russia. From the rugged Siberian steppes, through the expansive Asery Horizon formations found near St. Petersburg, to the soaring Ural Mountains that border Kazakhstan, this region's sedimentary layers seem to literally bulge with an ever-expanding variety of trilobite material. These include the midsized (4 centimeter) Carboniferous species *Ditomopyge producta*, which hails from Russia's Volgograd region, some 1,000 kilometers south of Moscow.

PILTONIA CARLAKERTISAE

Deep in the heart of New Mexico's rugged Caballo Mountains lies the Mississippian-age Lake Valley Formation. There, bands of gray- to pink-toned cherty limestone house an impressive array of well-preserved trilobite species, including *Piltonia carlakertisae*. Some 330 million years ago, these arthropods were all part of a rich marine faunal assemblage that included crinoids, corals, and brachiopods, many of which can also be found fossilized in this formation's sedimentary layers.

AMEROPILTONIA LAURADANAE

Although never common, the genus *Ameropiltonia* can be discovered in a variety of Mississippian-age locations across the face of North America. This relatively diminutive trilo-type, which generally runs between 1 and 3 centimeters in length, exhibits many of the "classic" features of the Proetida order, including a distinctly ovate body shape. The site that produces *Ameropiltonia* in Missouri's Choteau Formation is noted for yielding a relatively high percentage of complete specimens.

CUMMINGELLA CARRINGTONENSIS

Following a global extinction event that occurred at the end of the Devonian Period, by the beginning of the ensuing Mississippian, the entire trilobite lineage had been reduced to one surviving order. Examples of the rare proetid *Cummingella carringtonensis* from Castelton near Derbyshire, England, indicate how radically different—both in size and shape—late-stage trilobites were from many of their more primal ancestors. In most key morphological matters, however, these last-in-line representatives retained many of the quintessential qualities of the eminent trilobite class.

CUMMINGELLA BELISAMA

The Lower Mississippian trilobites found in the sedimentary outcrops that surround Tournai, Belgium, traditionally display a thin, white, mineralized coating. Some, however, including *Cummingella belisama*, occasionally manage to retain fossilized remnants of their calcified external shells. Once prevalent on the international natural history market, in recent years these 320-million-year-old trilobites have become scarce commodities, especially because many of the renowned Tournai Formation quarries have been permanently closed to paleontological exploration. Similar examples of this successful genus can be uncovered in Spain, Germany, and Ireland.

HESSLERIDES ARCENTENSIS

With each passing year, new trilobite localities are being brought to light and previously unknown species subsequently made available for academic analysis and commercial sale. Although material emanating from the Lake Valley Formation in New Mexico has been recognized and studied for decades, complete examples of *Hesslerides arcentensis* were first found—in extremely limited numbers—only during the early days of this century.

AMEROPILTONIA LAURADANAE (BREZINSKI, 2000)
Mississippian, Kinderhookian Stage; Chouteau Formation; Sedalia, Pettis County, Missouri, United States; Each specimen 1.8 cm

Despite continued exploration, these Mississippian-age specimens remain a rare and much coveted collectors' commodity.

HESSLERIDES BUFO

Hailing from the famed crinoid beds of Crawfordsville, Indiana, these interesting proetid trilobites—which could grow up to 6 centimeters in length but were usually half that size—seemed to coexist quite nicely in their chosen marine ecosystem, particularly prospering amid the swaying fields of crinoids for which the surrounding Edwardsville Formation is famous. By the way, for those wondering, crinoids are organisms closely related to starfish, and it is speculated that the symbiotic relationship often shared by crinoids and trilobites probably benefited both faunal components.

(TOP) *CUMMINGELLA CARRINGTONENSIS* (WOODWARD, 1884)

Lower Carboniferous (Mississippian); Dinantian Stage, Asbian Substage; Treak Cliff; Castelton, Derbyshire, England; 2.9 cm

(RIGHT) *PARTICEPS SP.*

Lower Carboniferous (Mississippian); Dinantian Stage, Brigantian Substage; Six Yard Limestone Series; "Ledge Marl" Member of Christon Bank; Alnwick, Northumberland, England

(LEFT) CUMMINGELLA BELISAMA HAHN, HAHN, AND BRAUCKMANN, 1985

Lower Carboniferous, Tournaisian (Mississippian); Tournai Formation; Tournai, Belgium; 2.2 cm

(BOTTOM) PHILLIPSIA (PHILLIPSIA) ORNATA BELGICA OSMÓLSKA, 1970

Lower Mississippian, Upper Tournaisian; Soignies Formation; Soignies, Belgium; 2.3 cm

HESSLERIDES ARCENTENSIS (HESSLER, 1962)

Lower Mississippian; Lake Valley Formation; New Mexico, United States; 3.2 cm

20

10 INTRIGUING
TRILOBITE QUESTIONS

Trilobites have been found just about everywhere that sedimentary outcrops of the proper age exist—from Morocco's imposing Atlas Mountains, to Bolivia's soaring Altiplano, to the majestic, snow-capped peaks of British Columbia, Canada. They have been uncovered on the desolate Siberian Steppes, on the shores of frigid Scandinavian islands, in the bustling suburbs of major North American cities, and in virtually every other Paleozoic place in between. From present-day England to Argentina, from Australia to China, from Greenland to the Czech Republic, the fossil record provides ample evidence that trilobites filled virtually every saltwater habitat in their primeval domain. In the Paleozoic, the roughly 290-million-year era that ran from the beginning of the Cambrian through the end of the Permian, the surface of our planet looked radically different than it does today. The continents we now instantly identify by their distinctive shapes and familiar global alignment were yet to be transported to their current lithospheric locations via the geological phenomenon of plate tectonics. Instead they often lay packed together in nondescript clusters in the Earth's southern hemisphere with one or at times two major landmasses dominating what was otherwise a water world. Significant sections of what we presently recognize as the continents of North America, South America, Asia, Africa, Australia, Europe, and even Antarctica were then under sea level, allowing marine forms such as trilobites to become fossilized in locations now hundreds of kilometers from the nearest prominent body of water. For all of these

PARABOLINA N. SP.
Upper Cambrian; Trinity Bay; Newfoundland, Canada; 6.2 cm

fossil record and may represent singularly distinctive examples of their kind, but others have been found by the thousands—or even the tens-of-thousands. Indeed, it has been estimated that more than 100,000 examples of the ubiquitous Middle Cambrian species *Elrathia kingii* have been unearthed in the Wheeler Formation outcrops of Utah.

HOW LONG DID A TRILOBITE LIVE?

As with many of the questions addressed in this section, we don't have a definitive answer regarding how long a trilobite lived. We can, however, extrapolate on the subject by using knowledge provided by contemporary members of the arthropod family. Lobsters, for example, are known to live for more than a century, and certain African termites are believed to live up to 60 years. It's not unreasonable to assume—especially when one considers the prodigious proportions some species attained—that certain trilobites also enjoyed lives that spanned decades.

reasons and many more, trilobites—including genera such as *Parabolina* and *Gabriceraurus*—have emerged as being among the most fascinating and coveted remnants of the planet's distant past. But for every question that science has answered about these incredible invertebrates, a new one seems to emerge.

Here are 10 intriguing trilobite questions along with a corresponding number of answers.

ARE ALL TRILOBITE FOSSILS RARE?

The simplest response to this popular query is no. Some trilobite species are exceedingly rare in the

WHAT DID TRILOBITES EAT?

As one of the top-of-the-food-chain classes in many of their marine ecosystems, trilobites most likely had their choice when it came to both their dining habits and the available items on their menu. It is speculated that some species used their hypostome (mouth plate) to scrape parasites off surrounding rock surfaces, and others may have been filter feeders, devouring early planktonic organisms by the mouthful. Some may have preyed on the numerous soft-bodied creatures that surrounded them in those primal seas. Still others may have been cannibalistic, feeding off the remains of deceased trilobites or even attacking live ones.

HOW DID TRILOBITES REPRODUCE?

Fossilized evidence of trilobite reproductive methods has long been considered among the Holy Grail subjects of paleontological research. The renowned paleontologist Charles Walcott spent many of his later years wading through the thousands of specimens he had collected in search of clues that might shed light on this age-old riddle. But it wasn't until the twenty-first century that unmistakable proof that trilobites laid eggs was provided by a series of beautifully preserved *Triarthrus eatoni* specimens found in the Ordovician outcrops of New York State. In addition, a 2022 report presented clear-cut evidence that even in the Cambrian male trilobites already possessed claspers, a common sexual feature in contemporary arthropods.

WHAT COLOR WERE TRILOBITES?

Trilobite fossils run the gamut from black, to brown, to red, to tan, to alabaster, though their real-life calcite carapaces may have been even more dramatic in their coloration. Much like tropical birds or exotic fish, it is not difficult to imagine swarms of brightly colored trilobites swimming or crawling through the primeval seas in search of food, shelter, or other members of their hard-shelled species. Unfortunately, despite a smattering of controversial academic evidence indicating that some trilobites may have presented a mottled or striped shell pattern to their marine world, we will probably never know a trilobite's true-life coloration.

DID TRILOBITES CRAWL OR SWIM?

They apparently did both. The scientific evidence is that different species of trilobites inhabited just about every available niche in their undersea domain. Some burrowed into the ooze that lay along the ocean floor. Others scurried along the bottom on rows of powerful walking legs. Still others used their fancifully constructed external morphology to ride the marine currents, and some proved to be powerful swimmers, using their hydrodynamically designed bodies to shoot through the sea's murky depths.

ARE TRILOBITE FOSSILS STILL BEING DISCOVERED?

Yes! Despite an ever-expanding spate of rules, regulations, and restrictions that have served to limit fossil exploration in certain corners of the globe in recent years, the fact is that if there's an arthropod-inspired will, there's apparently a way. And when it comes to discovering new trilobite sites, deftly excavating those locations, and properly preparing the material thus derived, there is more than a will—there is an almost uncontrollable passion. Whether it's a Lower Cambrian repository in the wilds of Alabama, an Ordovician outpost near a desert-hugging Moroccan village, or a Devonian deposit located high in the mountains of Bolivia, new trilobite locations—and new trilobite species—are still regularly being uncovered across the planet.

DID TRILOBITES VENTURE ONTO LAND?

Apparently they did venture onto land, albeit briefly, and they tended not to go very far in their travels beyond their protective marine ecosystem. Fossil finds from the Appalachian Mountains of Tennessee show that as far back as the Lower Cambrian certain species of *Olenellus* may have temporarily ventured onto neighboring tidal flats, perhaps in search of food or to lay their eggs. Trilobites were among the very first creatures to sojourn out of the sea. We know of this remarkable

GABRICERAURUS SP.

Upper Ordovician; Verulam Formation; Hastings County, Ontario, Canada; 9.2 cm

ZACANTHOIDES ROMINGERI
RESSER, 1942
**Middle Cambrian, Miaolingian Series, Wuliuan
Stage; Langston Formation, Spence Shale
Member; Box Elder and Cache counties, Utah,
United States; 8 cm**

development in the history of our planet because of evidence provided by the fossilized impressions (*Cruziana rugosa*) generated by their walking legs. Some of these impressions clearly appear to have been made on dry—or at least muddy—land.

ARE THERE ANY LIVING TRILOBITE DESCENDANTS?

Quite simply, it is now an irrefutable scientific fact that at the conclusion of the Permian Period some quarter of a billion years ago the entire trilobite class came to a rather abrupt and eternal end. Despite any lingering conjecture to the contrary—fortified by both the presence and the appearance of contemporary arthropods ranging from horseshoe crabs to pill bugs—after their incredible journey through more than 270 million years of evolutionary time, these highly resilient invertebrates left behind no direct living descendants.

WHO COLLECTS TRILOBITES?

Trilobite collecting recognizes no boundaries based on age, region, race, or religion. Whether you are a lawyer in Los Angeles, a banker in Beirut, or a factory worker in Frankfurt, you can find a way to collect these wonderful remnants of the distant past. These incredible invertebrates have been collected for centuries, with stories of their commercial distribution dating back to the nineteenth century in England and Bohemia (the modern-day Czech Republic). Since then, thousands of scientifically inclined enthusiasts around the globe have been drawn to the unique charms presented by these long-gone denizens of the Deep Time seas.

***OGYGINUS CORNDENSIS* (MURCHISON, 1839)**

Middle Ordovician; Llanvirn Series; Gilwern Hill, Hundred House; Builth Wells, Powys, Wales, United Kingdom; 8.5 cm

(LEFT) BATHYURISCUS FORMOSUS DEISS, 1939

Middle Cambrian; Meagher Formation; Horseshoe Hills; Manhattan, Montana, United States; 3.5 cm

(BOTTOM) EOCERAURUS TRAPEZOIDALIS (ESKER, 1964)

Upper Ordovician; Bromide Formation, Pooleville Member; Criner Hills, Oklahoma, United States; 6.2 cm

(LEFT) *CERAURUS SP.*

Middle Ordovician; Salona Formation; Salona, Pennsylvania, United States; 3.0 cm

(BOTTOM) *DALMANITES LIMULURUS LUNATUS (LAKE, 1904)*

Lower Silurian, Niagaran Series, Medina Stage; Edgewood Formation; Grafton, Illinois, United States; 5.8 cm

21

10 VERIFIABLY VENTRAL TRILOBITES

Even among the most astute trilobite collectors, it's relatively easy to confuse a ventral specimen with a negative example. These Paleozoic relics may initially look the same, but under closer inspection the differences between the two quickly become apparent. Negative, or counterpart, trilobites are exactly that—the reverse impression in the rock caused by the actual fossilized animal. Often these are left in the field by overzealous collectors who believe they are of little use, or value, to those looking for "real" trilobite fossils—although, truth be told, bits of outer shell, delicate genal spines, and even occasional compound eyes can remain affixed to a specimen's negative side. In contrast, ventral trilobites are complete examples of the actual fossils that have been preserved and subsequently discovered with the underside of their calcite carapaces showing. When properly prepared, these ventral samples may display delicate internal muscle attachment "hooks" and scars, along with appendage remnants positioned along the trilobite's underside. Many ventral specimens also showcase the trilobites' distinctively pronged hypostome, or mouth plate. These calcified, forked extensions (which may have allowed the trilobite to attach to a food source, rock surface, or mate) come in a wide variety of shapes and sizes, and when found independently can serve to identify the trilobite species from which they originated. Some locations, such as the famed Burgess Shale in British Columbia, the Hunsruck Slate of Germany, and the Lorraine Shale of upstate New York, have become renowned for producing trilobites in which ventral examples can provide detailed evidence of soft-tissue preservation,

ORYCTOCEPHALITES PALMERI (SUNDBERG AND McCOLLUM, 1997)

Lower Cambrian; Combined Metals Member, Pioche Shale; Panaca, Nevada, United States; 1.2 cm

number of intrepid adventurers and academics first started traversing through the area's rugged outcrops. Ironically, a complete dorsally preserved example of this species has yet to be discovered.

ARCTINURUS BOLTONI

When a complete ventral example of *Arctinurus boltoni* is unearthed in the famed Rochester Shale quarry of Middleport, New York, it is usually "flipped" during preparation to turn it into a more conventional—and marketable—dorsal specimen. But on occasion both ventral and dorsal examples of this once-rare species are discovered on the same Silurian-age matrix, resulting in a showpiece that beautifully exhibits both sides of trilobite morphology.

ISOTELUS GIGAS

Perhaps the most renowned and recognized trilobite to emerge from the Ordovician outcrops of eastern Canada and central New York State, *Isotelus gigas* is famous for its thick-shelled, three-dimensional preservation. Yet when a ventral version of these large, football-shaped asaphids is uncovered and properly prepared, the sheer "soup tureen" dimensions of these awe-inspiring trilobites come into even more dramatic focus.

including antennae, claws, gills, claspers, and even, on exceptionally rare occasions, eggs.

Here are 10 trilobites not shy about revealing their ventral sides.

ORYCTOCEPHALITES PALMERI

These diminutive (1- to 2-centimeter) Lower Cambrian trilobites are often dramatically preserved as color-contrasted ventral outlines. *Oryctocephalites* are found in a still rarely explored exposure of the now familiar Pioche Shale in Nevada, a fossil-filled site that was generally unknown—or at least overlooked—until the mid-1970s. At that time, a small

DALMANITES SP.

Examples of this unusual midwestern species—considered until recently a type of *Odontochile*—can appear fossilized in both dorsal and ventral forms on the same stone surface. This Paleozoic evidence suggests that their remains may have been part of a fast-evaporating tidal pool that left the resident trilobites quite literally high and dry. Also notable is the strange, inverted "groove" on the trilobite's pygidium, which may have aided this Silurian *Dalmanites* species during enrollment.

***ARCTINURUS BOLTONI* (BIGSBY, 1825)**

Lower Silurian; Rochester Shale Formation; Middleport Quarry; Middleport, New York, United States; ventral is 10.5 cm

RHENOPS LETHAEAE

The thin, coal-gray slate sheets that comprise the Devonian deposits of Hunsruck, Germany, often present specimens preserved in ventral relief. These attractive trilobites are most frequently of the phacopid *Chotecops ferdinandi.* But on rare occasions flipped examples of *Rhenops lethaeae* can be found—some tinged with pyrite—and when subjected to the proper degree of X-ray technology, they may also display aspects of that trilobite's ventral anatomy, including legs and gills.

ASAPHUS CORNUTUS

It's not that unusual for a well-prepared ventral trilobite specimen to have its fork-shaped hypostome exposed. This is a feature clearly visible on some examples of *Asaphus* drawn from the fossil-filled, Ordovician-age Asery Horizon levels that surround St. Petersburg, Russia. It has been scientifically speculated that this trilobite's hypostome was used to attach itself to a rocky surface during feeding, or possibly to a mate.

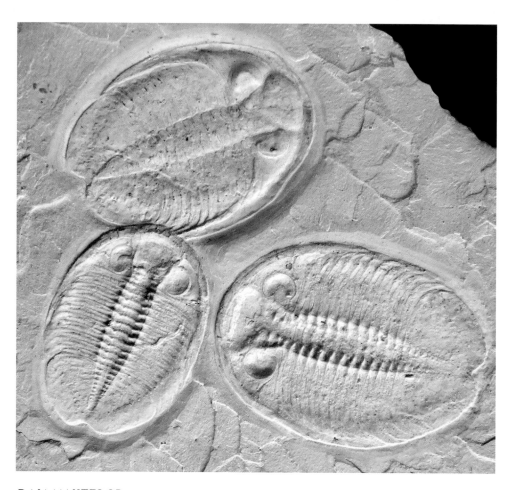

DALMANITES SP.
Silurian, Late Llandoverian—Early Wenlockian; Joliet Formation, Brandon Bridge Member; Waukesha County, Wisconsin, United States; largest trilobite 4 cm

CYPHOPROETUS SP.

This represents the only known ventral example of *Cyphoproetus* found in the renowned Lorraine Shale trilobite beds of New York State, a site that has previously produced hundreds of beautifully preserved *Triarthrus eatoni* specimens. This legendary outcrop's small but dramatically fossilized trilobites are often uncovered ventrally, with golden pyrite replacing the creature's original calcite morphology. For more than a century, science has recognized both the unique preservation displayed by these Ordovician organisms and the fact that many of those ventral examples routinely reveal a stunning panoply of trilobite soft-body parts, including various appendages and antennae, which are clearly visible here.

CERAURUS PLATTINENSIS

In certain North American localities, entire layers of *Ceraurus* trilobites have been preserved ventrally. It has been scientifically speculated that much like modern horseshoe crabs, some species of trilobites—including *Ceraurus plattinensis*—may have swum upside down during their time in the primal seas. Thus finding such species preserved ventrally in sedimentary bedding planes isn't particularly surprising.

OLENELLUS FOWLERI

The legendary Lower Cambrian trilobite beds of the western United States are often figuratively brimming with faunal elements. Thickly banded layers in famous formations such as the Pioche Shale of Nevada can be found covered with trilobite debris—including complete and colorful examples of species such as *Olenellus fowleri*. In some special cases, the trilobite's thick calcite shell will adhere to its "negative" side, thus making it a true and truly attractive ventral specimen.

LICHAS MAROCANUS

As demonstrated on some of the magnificently preserved fossils that emerge from the North African Lagerstatte, at times dorsal and ventral trilobite specimens can appear on the same sedimentary stone. Here a large (11 centimeter) example of the rare Ordovician species *Lichas marocanus* reveals that natural phenomenon, with the trilobite being presented in ventral relief alongside two dorsal specimens. It is speculated that such unusual preservation may be the result of one of the fossils being flipped during a storm, a time when members of a trilobite species may have gathered together for mutual protection.

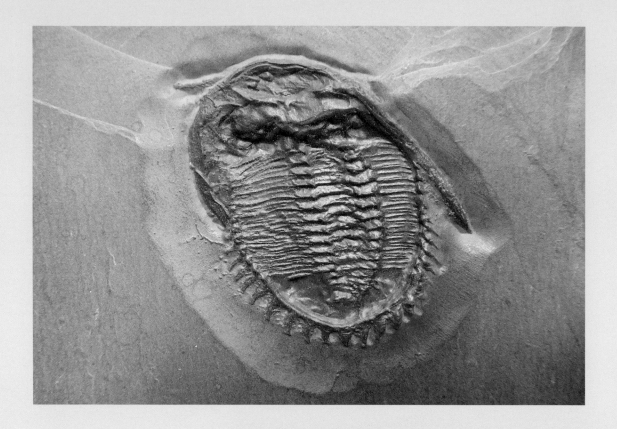

**(TOP) *RHENOPS LETHAEAE*
(KAYSER, 1889)**

**Lower Devonian, Seigenian/Emsian Stage;
Hunsruck Slate; Bundenbach, Germany;
5.6 cm**

(RIGHT) *CYPHOPROETUS SP.*

**Upper Ordovician; Lorraine Shale, Martin
Quarry; Beecher's Trilobite Bed, Oneida
County; New York, United States; 1.6 cm**

Photo courtesy of Markus Martin

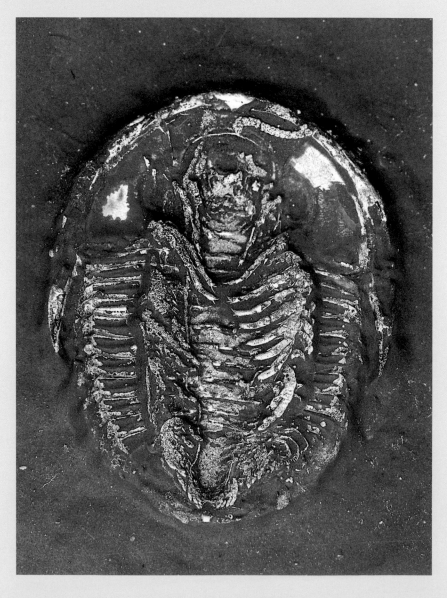

(LEFT) *ASAPHUS CORNUTUS* PANDER, 1830

Middle Ordovician; Aseri Regional Stage; Duboviki Formation; Vilpovitsky Quarry, St. Petersburg region, Russia; 9.5 cm

(BOTTOM) *OLENELLUS FOWLERI* (PALMER, 1998)

Lower Cambrian; Pioche Shale Formation; Nevada, United States; 6.5 cm

LICHAS MAROCANUS (DESTOMBES, 1968)

Middle Ordovician; Ktaoua Formation; Tazarine, Morocco; Ventral (top) is 11.3 cm

22

10 ROCK 'N' ROLLER TRILOBITES

With their thick, sometimes spine-encrusted calcite shells, an enrolled trilobite could present quite a daunting challenge to any marine predator seeking an easy meal. During enrollment, when these primal arthropods assumed a strikingly symmetrical, ball-like shape, the flexibility provided by their thoracic segments enabled the trilobite's rigid cephalon and pygidium to interlock. Assuming such a compact stance encased the creature's delicate organs and soft ventral appendages, making the trilobite virtually impervious to direct attack from all but the most intimidating of undersea bullies. The fossil record indicates that a preponderance of the 25,000 recognized trilobite species could enroll and that this practice was common among virtually all orders of these unique invertebrates. This feature lasted until the demise of the trilobite line at the end of the Permian, but when did this notable evolutionary development begin? A few Cambrian trilobites appear to have possessed the ability to enroll, although the rare fossilized remains of these specimens seem more "folded" than truly enrolled. But by the Lower Ordovician, some 480 million years ago, trilobite enrollment was already quite advanced. Many of the magnificent specimens emerging from Russia's famed Ordovician-age Asery Horizon quarries are found in various states of enrollment, frequently presenting the appearance of nearly perfect fossil spheres. By the Devonian, perhaps the most renowned of these trilobite "rollers" had emerged. These are the beautifully preserved *Eldredgeops* species from Ohio, where their 400-million-year-old circular conformity is broken only by the pronounced protrusions made by their compound eyes. Indeed, judging by all the fossil evidence, it is apparent that the ability to enroll played a vital role in helping trilobites not only to survive but to flourish, allowing them to dominate the planet's seas for more than a quarter-of- a-billion years.

Here are 10 notably enrolled trilobites.

DIPLEURA DEKAYI

When examples of these large homalonotids were first uncovered across the northeastern United States in the late nineteenth century, their uniformly enrolled shape confused many of the academics that confronted them. In fact, as strange as it may now seem, it was only after numerous outstretched examples had been subsequently unearthed that scientists realized that not all *Dipleura dekayi* were enrolled!

PRESBYNELIUS IBEXENSIS

To field collectors, one of the appealing qualities exhibited by many enrolled trilobites is that their mode of preservation often allows them to be more easily freed from their surrounding matrix. Small examples of the rare species *Presbynelius ibexensis* from the Ordovician sediments of Utah (which are usually no more than 5 centimeters in diameter) can even be found lying loose amid the scattered scree, just waiting to be picked up by some fortunate fossil-seeking explorer.

DIPLEURA DEKAYI (GREEN, 1832)
Middle Devonian, Givetian; Skaneateles Formation, Hamilton Group; Delphi Station Member, New York, United States; diameter: 6.6 cm

ELDREDGEOPS CRASSITUBERCULATA

One of the most famous trilobite species in the paleontological world is also among its most renowned "rollers." Often found as perfectly symmetrical calcite-covered spheres—with only their prominent eyes breaking their circular conformity—enrolled examples of the *Eldredgeops* genus (best known from the Devonian outcrops of Ohio, New York, and Michigan) are expected components of any private or public trilobite assemblage.

FLEXICALYMENE RETRORSA

There's good reason for these common midwestern trilobites to have "flex" as part of their generic nomenclature—the ability of *Flexicalymene* to bend their thorax during enrollment is perhaps the species' most distinguishing morphological characteristic. These Ordovician bugs are so pervasive in some outcrops that lucky collectors in Ohio, Indiana, and Kentucky often possess shelves packed with scores of sediment-free flexi rollers they've gathered over the years.

PTERYGOMETOPUS ANGULATUS

The fossil-rich deposits of western Russia's famed Asery Horizon have produced dozens of Ordovician-age trilobite species. Some, such as the attractive asaphid *Pterygometopus angulatus*, display an almost unmatched ability to enroll. In fact, many of the species from this locality are found in a tightly enrolled position, and academia is currently hard at work attempting to decipher the half-a-billion-old riddles presented by such a unique morphological stance, particularly whether it reflects a defensive or possibly a postmortem posture.

CAUDILLAENUS NICOLASI

For nearly half a century the trilobites of Morocco have amazed, inspired, and entertained collectors around the globe. Among the most coveted of these Paleozoic prizes have long been examples of large rollers such as those presented by the relatively recent discovery of *Caudillaenus nicolasi*, an unusually shaped Ordovician species first described in scientific literature in 2014. It is interesting that the 20-odd examples of this trilobite with an elongated pygidium found so far have all emerged from a single quarry located 18 kilometers due south of the fossil-trading hotbed of Alnif, and more than half of these are enrolled.

PARAHOMALONOTUS CALVUS

It is sometimes mistakenly surmised that only relatively small species of trilobites needed to enroll for protection. The fact is that a variety of large genera—including diverse asaphids, phacopids, and homalonotids—assumed such a defensive stance during their time in the primal seas. Few examples, however, demonstrate this characteristic more dramatically than *Parahomalonotus calvus* from the Devonian of Morocco. Indeed, those who have seen these specimens—which can exceed 10 centimeters in diameter—jocularly refer to them as "baseball bugs" due to their size and circular configuration.

ACERNASPIS SP.

On very rare occasions, enrolled trilobites can become fossilized as translucent hollow crystals. When these unusual specimens are properly (and delicately) separated from any surrounding matrix and subjected to a strong white backlight, their cleaned carapaces can ignite with an almost incendiary glow. This bright orange,

**ELDREDGEOPS
CRASSITUBERCULATA
(STUMM, 1953)**

Middle Devonian, Givetian; Silica Shale
Formation; Sylvania, Lucas County,
Ohio, United States; diameter: 2.9 cm

440-million-year-old Silurian-age specimen from Quebec displays a crystal-clear, "gem" quality preservation that may be unique to trilobites hailing from this remote locality.

NILEUS ARMADILLO

Perhaps no European trilobite species represents the morphological phenomenon of enrollment better than the aptly named *Nileus armadillo*. So many of these midsized arthropods (up to 4 centimeters in diameter) have been discovered in an enrolled position throughout the Ordovician outcrops of Norway and Sweden that the entire species has been named after the creature that is

perhaps the most renowned modern-day proponent of enrolling, the armadillo.

HOLMIELLA FAIX

This unusual specimen is included here not because it displays a great example of trilobite enrollment but because it could well represent the *first* documented case of such behavior exhibited in the fossil record. Indeed, some 520 million years ago, soon after trilobites initially began filling the Cambrian oceans, species such as *Holmiella faix* may have attempted to roll up their segmented calcite carapaces to protect themselves from predators lurking in those increasingly savage seas.

CAUDILLAENUS NICOLASI RABANO, GUTIERREZ-MARCO, AND GARCIA-BULLIDO, 2014

Middle Ordovician; Taddrist Formation; Jebal Rahiat; Taychoute, Morocco; diameter: 6.7 cm

PARAHOMALONOTUS CALVUS CHATTERTON, FORTEY, BRETT, GIBB, AND McKELLAR, 2006

Middle Devonian; Timrhanrhart Formation; Jibel Gara el Zguilma, Morocco; diameter: 10 cm

(BOTTOM) ***ACERNASPIS SP.***

Early Silurian, Llandovery, Early Telychian; Jupiter Formation, Ferrum Member; Anticosti Island, Quebec, Canada; diameter: 1.5 cm. This is a translucent, crystalized trilobite that has been backlit.

Collage courtesy of Markus Martin

HOLMIELLA FAIX HOLLINGSWORTH, 2006

Lower Cambrian; Poleta Formation; Montezuma Range; Nevada, United States; 3 cm across genal spines

23

10 PERTINENT PREPARATION STEPS

In the third decade of the twenty-first century, the art of trilobite preparation has risen to never before imagined levels of dexterity, creativity, and accomplishment. Ironically, centuries ago one of the world's most renowned artists explained his approach to sculpture as merely freeing already-existing shapes from their surrounding stone coverings. When we shift our focus to those who prepare trilobites, that method of extraction describes *exactly* how it's done. Perhaps those who deal with fossils don't possess quite the inherent level of artistic elan exhibited by a Renaissance master, but when all is said and done, it is trilobite preparators who truly free existing forms from their encasing stone matrices. Yet for all of their apparent skill, the deft work done by trilobite preparators has only been properly acknowledged in relatively recent times. Over the last three-plus decades in particular, things have changed radically on the trilo-prep scene. If we venture back to the late 1980s, fossil preparation was still basically an arduous, often haphazard process. Acid baths, wire brushes, and handheld dental tools persistently left the resulting trilobites looking like bruised and battered remnants. By the dawn of this century, however, prep work had evolved into a state-of-the-art procedure that employs an exotic assortment of pneumatic scribes, air abrasive machines, and powerful vacuum ventilators. In the proper hands, these devices frequently prove capable of miraculously transforming even the most derelict half-a-billion-year-old trilobite into a nearly flawless Paleozoic prize—one guaranteed to garner admiration from both the academic and the collecting communities.

Here is a list of 10 pertinent preparation steps.

ISOTELUS WALCOTTI WALCOTT, 1918
Upper Ordovician; Bobcaygeon Formation; Ontario, Canada; largest specimen 7.2 cm

RECOGNIZE A COMPLETE SPECIMEN

Whether in the field or at a local rock and mineral show, a highly trained eye is required to differentiate the partially exposed remnants of a complete trilobite specimen from the tantalizing residue of a fossilized fragment. No preparator wants to waste time, money, and energy only to subsequently discover that the trilobite is only half there.

IDENTIFY THE SPECIES

Knowing the species (or at least the genus) of the trilobite upon which you're working will greatly aid in all prep-associated activities. Sometimes a specimen emerges from its sedimentary strata in cross section; other times it breaks along the bedding plane. This information should provide anyone with a keen eye an essential clue when it comes to the best means of preparing the trilobite before them.

MAKE SURE YOU USE THE PROPER EQUIPMENT

Modern technology has provided preparators with an amazing assortment of devices—including air scribes, drills, and abrasive machines—to aid in the performance of their self-appointed trilobite-related task. Long gone are the days when dull, hand-held dental tools were all prep artists possessed in their arthropod cleaning arsenal. Today's trilobites clearly benefit from this high-tech improvement. Quite simply, the better your equipment, the better the resulting specimen should be.

VENTILATION IS VITAL

Both the powder used during the abrasive cleaning process and the resulting silica-laden dust that is often an inherent ingredient of a trilobite's Paleozoic matrix are dangerous to inhale. Proper masks and well-ventilated prep cabinets are a necessity for those who wish to undertake the preparation procedure. Be assured that the long-term effects can be dire if you don't heed these precautions.

BE PATIENT

By their nature, most trilobite enthusiasts tend to be an impatient bunch—they want their specimens, and they want them now! In direct contrast, those who practice the preparation art need to be calm of character, deft of hand, and stoic of spirit. Much like slowly baking a soufflé, when the subject is trilobites, patience and proper procedure are key ingredients to achieving a desired preparation result.

URALICHAS SP.
Lower Ordovician; Fezouata Formation; Drâa Valley, Zagora, Morocco; 6.3 cm

It is wise to recognize whether your preparation strengths lie in being a skilled technician or a true artist—and know that *both* qualities can be of equal and essential value.

BE CAUTIOUS WITH SPINES

Many of today's perfectly prepared "next generation" trilobites feature scores of delicate free-standing spines—some no thicker than a strand of angel hair pasta—along with amazingly complex compound eyes, and even occasional soft-part preservation. These morphological features make these specimens both a joy to observe and a pleasure to study, but they are a pain in the pygidium to prep.

CREATE A PROPER BASE

For many collectors, the presentation of their prized, perfectly prepped trilobite is almost as important as the specimen itself. A resourceful preparator should always remember to either create a flat base of matrix upon which the trilobite can rest during display or leave at least one level edge that will securely fit within the dimensions of a sturdy plastic or wooden stand.

PACK CAREFULLY

Especially for those who prepare specimens for fellow collectors, it is essential that the final step—that of properly packaging the specimen for shipment—is executed with as much skill and care as any other step in the prep process. Many a magnificent trilobite has suffered a disastrous fate after being improperly secured prior to hitting the postal system maelstrom. Some shipments from Morocco and Russia even use screws or putty to properly stabilize their delicate Paleozoic cargo.

BE PREPARED FOR ACCIDENTS

No matter how careful you may be, accidents do happen during the preparation process. Sometimes bits of trilobite shell flake off during cleaning, or an entire spine may be broken by a slightly overzealous stroke. Various glues and epoxies should be constantly on hand, with each available to quickly and effectively turn a major problem into a minor (and expected) distraction.

ARE YOU A TECHNICIAN OR AN ARTIST?

Not every preparator can exhibit the artistic skills of a Renaissance master, but some among us do inherently possess more creative flair than others.

GIGANTOPYGUS N. SP.

Lower Cambrian, Atabanian Stage; *Neltneria* Layer; Issafen Formation; Issafen, Morocco; 12.3 cm

(LEFT) *OLENELLUS TERMINATUS PALMER*, 1998

Lower Cambrian; Combined Metals Member; Pioche Shale, between Klondike Gap and Ruin Wash faunas; Nevada, United States; 5.3 cm

(BOTTOM) *JUJUYASPIS KEIDELI KOBAYASHI*, 1936

Lower Ordovician; Orenburg region, Russia; Larger trilobite 2.7 cm

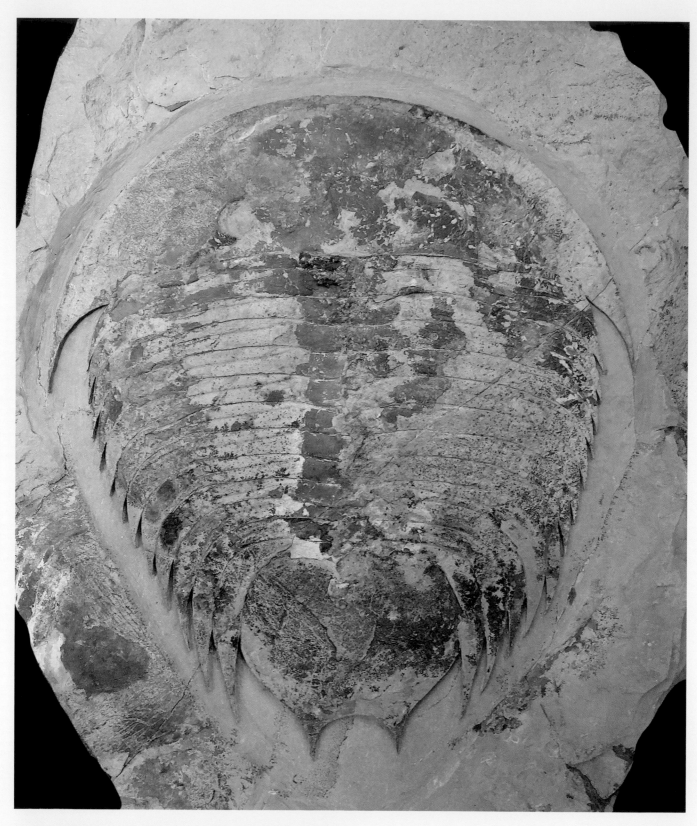

DIKELOKEPHALINA BRENCHLEYI (FORTEY, 2010)

Lower Ordovician; Fezouata Formation; Draa Valley, Morocco; 26.7 cm

24

10 CLEVERLY
COLORED TRILOBITES

The fossilized forms of trilobites display an amazing diversity of color. Depending on where they've been found and what minerals have most influenced their process of preservation, the rock-hard carapaces of these Paleozoic relics can appear in tones of black, white, beige, green, brown, red, orange, or yellow. From the golden hues that distinguish the myriad species unearthed in the Ordovician layers surrounding Russia's Volkhov River, to the ghostly alabaster preservation that defines much of Portugal's Valongo Formation, to the charcoal-gray that characterizes the Silurian material drawn from upstate New York's Rochester Shale, virtually every color contained in our planet's natural palette can be seen in the calcified remains of trilobite exoskeletons. Much like modern birds, butterflies, or tropical fish, it's not difficult to imagine trilobites of different genera swimming through those antediluvian oceans exhibiting a wide variety of contrasting tones. Indeed, distinctive external coloration appears to be a recurrent characteristic in the animal kingdom—and what is true today seems only logical to have been true when trilobites crawled through the seas some half-a-billion years ago. We know little about the color ornamentation of trilobite exoskeletons for a very good reason. Hundreds of millions of years after their demise, the various calcite-infused pigments that created those colors in life have long since been replaced by the invading minerals essential to the fossilization process. Although many trilobites found adorning private collections and museum shelves (including *Glyptambon verrucosus*, *Luciaspis matiasi*, and *Yinites yunnanensis*) often display a veritable rainbow's range of

SELENOPELTIS SP.

Lower Ordovician, Arenigian; Upper
Fezouata Formation; Zagora, Morocco;
3.2 cm

colors, it is unlikely that any of those shades represent the trilobite's actual hue during its life in the ancient seas.

Here's a look at 10 particularly colorful trilobites.

SELENOPELTIS SP.

Over the last decade, Morocco's Lower Ordovician Fezouata Formation has gained a rapidly expanding reputation among collectors, both for the unusual trilobite species it has produced (some featuring soft-part preservation) and for the equally distinctive colors that many of those specimens exude. Due to the unusual array of minerals present in the region, Fezouata trilobites, such as the rare *Selenopeltis* displayed here, may exhibit brilliant bursts of red, orange, and even the occasional purple tinge in their "autumnal" Paleozoic presentation.

BUMASTOIDES BECKERI

Many of the Ordovician trilobites that hail from the American Midwest, including *Bumastoides beckeri*, have been uncovered in long-studied formations such as the Galena, Maquoketa, and Prosser. These *Bumastoides* specimens possess a beautiful golden-hued calcite carapace that is particularly notable when contrasted with the charcoal-gray color that most members of the same genera (such as *Bumastoides holei* from the famed Walcott/Rust Quarry) display in other parts of the North American fossil landscape.

ACIDASPIS CINCINNATIENSIS

As evidenced by *Acidaspis cincinnatiensis* from the Ordovician-age Martinsburg Formation outcrops

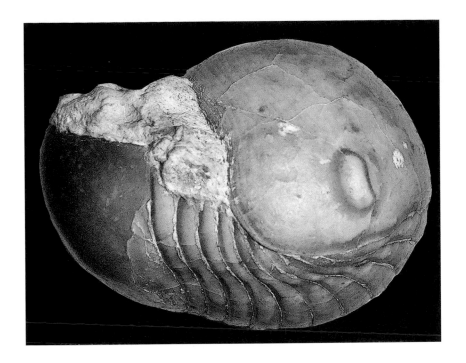

***BUMASTOIDES BECKERI
(SLOCUM, 1913)***

**Upper Ordovician; Maquoketa Formation,
Elgin Member; Fayette County, Iowa, United
States; 7.1 cm around the curve**

of eastern Pennsylvania, even without their distinctive calcite shells many trilobite species still manage to display a veritable panoply of colors in their fossilized remains—including shades of red, yellow, and orange. Often this appearance is caused by a thin coating of mineral replacement—such as limonite (iron oxide)—adhering to the trilobite's outer layers during the later stages of the fossilization process.

YINITES YUNNANENSIS

In the fossil world, white trilobites are about as unusual as the mythical white buffalo—you may see only a precious few in an entire lifetime of searching. Until the Lower Cambrian trilobite *Yinites* emerged from southern China in very small numbers during the second decade of the twenty-first century, even the most avid collector would have been hard-pressed to *name* a similarly toned species, especially one that had managed to retain fossilized remnants of its calcite shell.

CYBELELLA REX

In recent years, new quarries have continued to open in the fossil-rich Ordovician outcrops that surround St. Petersburg, Russia. Many of those repositories are yielding previously unknown species, some featuring a beautiful ivory-colored calcite exoskeleton. The spinose, 3-centimeter-long *Cybelella rex* represents one of the more ornate examples to emerge from this often explored area—a trilobite that takes more than 50 hours to painstakingly prepare.

LOCHMANOLENELLUS TRAPEZOIDALIS

On rare occasions, the fossilization process takes an unexpected twist and creates a multicolored trilobite. Such is sometimes the case with the rare and notably large (up to 25-centimeter-long) species *Lochmanolenellus trapezoidalis* from the Lower Cambrian Poleta Formation of Nevada. In some early trilobite species, the calcite shell adorning their cephalic region was apparently considerably thicker than the covering atop the remainder of their carapace (perhaps after a molt), which resulted in occasionally producing a veritable "pinto" pattern on their fossilized forms.

GLYPTAMBON VERRUCOSUS

The Waldron Shale outcrops of southern Indiana are globally renowned for the magnificent preservation of their fossilized fauna. Among the keystone trilobites to emerge from that 425-million-year-old Silurian site are large (up to 13 centimeter) examples of the dalmanitid *Glyptambon verrucosus*, which frequently present a mottled brown-and-black shell coloration that ranks among the more recognizable features on the U.S. paleontological scene.

LUCIASPIS MATIASI

In Spain's beautiful Cantabrian Mountains lurks the little explored Lancara Formation, a half-billion-year-old horizon that has been yielding examples of large, colorful trilobites such as *Luciaspis matiasi* over the last two decades. With its uniquely shaped pygidium and row of prominent axial spines, this rare Lower Cambrian species—which features preservation presenting tones of brown, tan, and black—has fast become a collector's favorite.

HUNGIOIDES BOHEMICUS AROUQUENSIS

The Valongo Formation of Portugal is renowned for the impressive sizes its fossilized trilobite specimens could attain—some, such as *Hungioides*, were up to 50 centimeters in length. But perhaps equally as intriguing as their oversized dimensions is the ghostly alabaster preservation that many of these Ordovician-age monsters present. That color contrasts rather dramatically against the charcoal-gray tone of the surrounding sedimentary stone, a fact that provides these eye-catching specimens with a hauntingly emotive aura.

TRIARTHRUS EATONI

Best known for their incredible soft-tissue preservation, which can include gills, legs, antennae, and even eggs, the *Triarthrus eatoni* specimens that emerge from the famed Beecher's Trilobite Bed quarry in upstate New York are perhaps equally famous for their magnificent pyritized preservation. With their golden shells (and soft-body parts) contrasting against the location's jet-black shale, these specimens are some of the most attractive and coveted trilobites in the entire world.

(LEFT) *ACIDASPIS CINCINNATIENSIS* **(MEEK, 1873)**

Upper Ordovician; Martinsburg Formation; Lebanon, Pennsylvania, United States; 2.5 cm

(BOTTOM) *YINITES YUNNANENSIS* **ZHANG W, 1966**

Lower Cambrian, Series 2, Stage 3; Qiongzhusian Regional Stage; Hongjingshao Formation; Southeastern Yunnan Province, China; 12.1 cm

LOCHMANOLENELLUS TRAPEZOIDALIS WEBSTER AND BOHACH, 2014

Lower Cambrian, Series 2, Lower Dyeran Regional Stage; Middle Member of the Poleta Formation; Montezuma Mountains,; Esmeralda County, Nevada, United States; 14.3 cm

***GLYPTAMBON VERRUCOSUS* (HALL, 1864)**

Middle Silurian; Waldron Shale Formation; Shelby County, Indiana, United States; 7.4 cm

LUCIASPIS MATIASI ÁLVARO, ESTEVE, GRACIA, AND ZAMORA, 2019

Lower Cambrian; Láncara Formation; Cantabrian Mountains, León Province, Spain; 10.1 cm

25

10 ENGAGINGLY EYED TRILOBITES

The first members of the trilobite class appeared in the world's oceans some 521 million years ago. By that time in Earth's history, these arthropods already featured highly developed mineralized eyes. This monumental evolutionary event marked trilobites as among the first creatures in the planet's 4.5-billion-year existence capable of leaving behind fossilized evidence of such a pronounced anatomical advance. Those original trilobite eyes were crescent-shaped, providing nearly 360-degree vision for the primitive members of the Redlichiida order that exhibited them—a major help when it came to keeping watch on the menacing cast of marine predators that surrounded them in those Lower Cambrian waters. Unlike any modern eyes, *all* trilobite eyes were constructed totally of calcite, providing these ancient creatures with a unique ability to perceive subtle changes in their aquatic environment. Scientists continue to be fascinated with the revolutionary logistics behind the development of these amazing mineralized ocular outlets, but trilobite collectors prefer to revel more in their singularly distinctive aesthetic. Some of those eyes, including the holochroal optics attached to the Ordovician *Asaphus kowalewskii* and the Silurian *Ommokris vigilans*, sat atop stalks up to 4 centimeters in length. Other eyes, especially during the Devonian, were perched along the trilobite's gently curving cephalon in a fashion most resembling the headlights of some fancy twenty-first-century European sports car. Phacopid species such as *Drotops armatus* and *Eldredgeops milleri* presented schizochroal compound eyes featuring dozens of geometric lenses stacked in rigid uniform rows. Indeed,

**OLENELLUS NEVADENSIS
(WALCOTT, 1910)**

Lower Cambrian; Pioche Formation,
Delamar Member; Klondike Gap;
Nevada, United States; 5.8 cm

when we stare into the fossilized eyes of a trilobite today, it takes only minimal imagination to sense that these Paleozoic animals may be looking back at us, providing a dramatic link to life some 500 million years in the past.

Here's some insight into 10 engagingly eyed trilobites.

OLENELLUS NEVADENSIS

The first trilobite eyes that appear in the fossil record—similar in both shape and structure to those attached to early species such as *Olenellus nevadensis*—were already both anatomically well-formed and evolutionarily well-advanced. In fact, recent Lower Cambrian discoveries made in Australia's Big Gully Formation (part of the Emu Bay Shale) indicate that arthropod eyes may have evolved as much as 10 million years earlier than the trilobites' characteristic outer shells.

LEVICERAURUS MAMMILOIDES

The eyes on this graceful Ordovician species found in the fossil-rich outcrops of eastern Canada are small and slightly raised. Some recent scientific speculation has indicated that such a shape may suggest that these are the eyes of a swimmer rather than a crawler. If true, this would provide unique insight not only into this highly collectible species' preferred mode of travel but also into trilobite behavior in those primal seas some 450 million years ago.

ELDREDGEOPS MILLERI

Perhaps no eyes in the fossil domain are as renowned as those attached to members of the *Eldredgeops* genus—named in honor of the noted American paleontologist Niles Eldredge. These intricately shaped (and often superbly preserved) ocular outlets sat atop this trilobite's gently curving cephalon in a manner that would have provided its host with an unparalleled view of the Devonian marine world around it. Known primarily from the famous Silica Shale outcrops of Ohio, *Eldredgeops* eyes rank among the most widely recognized—and often studied—features in the entire paleontological field.

OMMOKRIS VIGILANS

As unusual as some trilobite eyes may appear, few can match the midwestern species *Ommokris vigilans* in terms of sheer, out-of-this-world strangeness. With their ocular attachments perched atop centimeter-long stalks of connective calcite (here replaced by dolomite), few other trilobites have ever been more morphologically adapted to a unique and perhaps partially covered undersea lifestyle than this rare Silurian species.

DROTOPS MEGALOMANICUS

Among the more notable trilobites to emerge from the abundant Devonian deposits that line the Moroccan landscape, *Drotops megalomanicus* initially seemed merely to be a larger version of the various phacopid species that collectors already knew—and coveted—from locations found throughout North America. To a certain extent, that's exactly what it was. But the distinctive size and shape of the

lenses aligned in its huge schizochroal eyes clearly indicated that this species—which could be up to 16 centimeters in length—was decidedly different from any other member of its trilobite order.

FENESTRASPIS AMAUTA

An often debated topic among trilobite enthusiasts revolves around the question of which species possesses the most impressive eyes in the fossil record. The answer to that query may very well be the rare South American phacopid *Fenestraspis amauta*. The compound eyes on this Devonian species are simply huge! Some examples feature hundreds of lenses attached in tight rows to thick, 3-centimeter-high eye columns.

ERBENOCHILE ERBENI

If *Fenestraspis* didn't possess the most pronounced compound eyes in the trilobite realm, that distinction may well have fallen upon another Devonian-age phacopid, *Erbenochile erbeni*. Not only does this Moroccan species feature thick, towering, 2- to 3-centimeter-high schizochroal eye stacks—each topped by a unique shading brim—but it also has a row of intimidating defensive spines that extend along its axial lobe.

SELENECEME ACUTICAUDATA

At the opposite end of the extravagantly eyed arthropod spectrum featured here lurk such "blind" trilobites as the Ordovician-age *Seleneceme*, a genus that possessed no discernable eyes at all. Such an unexpected morphological profile indicates that these creatures lived at considerable marine depths where filtered or severely limited sunlight cast minimal illumination on their daily activities, thus negating their ocular needs. In fact, this eyeless evolutionary trend began quite early in the trilobites'

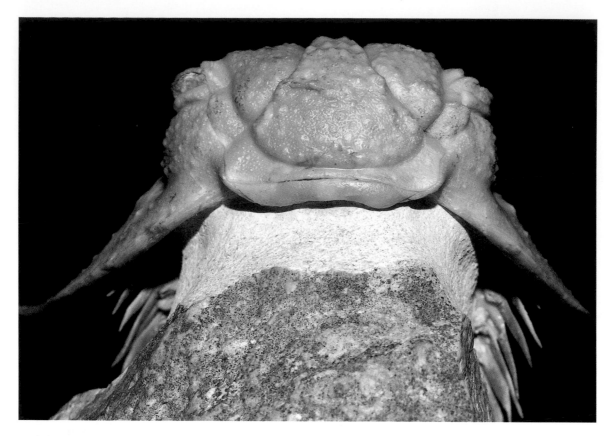

HOPLOLICHOIDES CONICOTUBERCULATUS (NIESZKOWSKI, 1859)

Upper Ordovician, Early Caradocian; Kukruse Regional Stage; Viivikonna Formation; Alekseevka Quarry; St. Petersburg region, Russia; 6.5 cm

Note: No fossilized eye lenses—accurate for most Russian lichids.

crawl through Deep Time, as exhibited by Middle Cambrian species such as *Conocoryphe sulzeri.*

ASAPHUS KOWALEWSKII

When this amazing Ordovician species from the fossil-filled deposits of western Russia first began appearing on the world stage in the mid-1990s, human eyes popped and jaws dropped. No one in the collecting community had ever seen anything quite like *Asaphus kowalewskii*, a trilobite with eyes perched atop 5-centimeter-long calcite stalks—a major evolutionary advantage for a species that may have lived most of its life partially buried in loose seafloor sediment.

SYMPHYSOPS STEVANINAE

The unusual Ordovician genus *Symphysops* is perhaps best known from the occasional discovery of fully articulated specimens in both Morocco and the Czech Republic. These trilobites can display a pair of huge, wraparound holochroal eyes featuring hundreds of small, tightly packed lenses that dominate the ventral side of the creature's cephalon. In the few complete examples that are known, these eyes actually merge to form a single cyclops-like ocular surface, a design that gave these free-floating trilobites an excellent and perhaps panoramic view of the seafloor beneath them.

***LEVICERAURUS MAMMILOIDES* (BARTON, 1913)**

Upper Ordovician; Cobourg Formation; Colbourne, Ontario, Canada; 6.1 cm

OMMOKRIS VIGILANS (HALL, 1861)

Lower Silurian, Niagaran Series, Medina Stage; Edgewood Formation; Grafton, Illinois, United States; 4 cm

SELENECEME ACUTICAUDATA (HICKS, 1875)
Lower Ordovician; Llanvirnian, Abereiddian Substage; Shelve Formation; Shropshire, England

SYMPHYSOPS STEVANINAE LÓPEZ-SORIANO AND CORBACHO, 2012

Upper Ordovician; Upper part of Lower Ktaoua Formation; Jebel Tizi n'Mouri, El Kaid Errami, Morocco; 5.5 cm

26

10 WAYS TO SPOT
A FAKE TRILOBITE

Over the last 30-plus years, a thriving and apparently still expanding side industry has grown up around trilobites. Native craftspeople, often working in rural villages without even the benefit of electricity, basically create their own brand of highly realistic fake fossils from mud, plastic, rubber, or just about any other reliably pliable compound upon which they can lay their artistically inclined hands. Such unsavory practices have long been an accepted or at least acknowledged part of some trilobite transactions, especially those stemming from the paleontological stronghold of Morocco. That nation's fossil trade first blossomed on the international scene in the late 1980s, and North African artisans have since become notorious for utilizing everything from local mud and mortar to automobile repair putty to carefully construct a veritable smorgasbord of faux trilobites. These "fossils" range from what may initially appear to be large Middle Cambrian *Paradoxides* and *Cambropallas* specimens to a mind-numbing variety of delicately barbed Devonian trilo-types. The work of these fabricators is so compelling—with their original molds being crafted from top-grade authentic examples—that even university-trained trilobite experts sometimes have been fooled. As many trilobite collectors have sadly come to learn, however, Morocco is far from the only place on planet Earth where such paleontological tomfoolery is taking place. In recent years, a similar but somewhat more sophisticated method of fossil fabrication has been established by those diggers, preppers, and merchants who operate in and around St. Petersburg, Russia. Local workers—many of whom also actively search the neighboring Ordovician quarries for authentic fossil material—have begun producing a line of trilobites that have been augmented, if not totally constructed, through the inventive use of high-tech polymer plastics and resins.

Here are 10 reliable ways to spot a fake trilobite.

AIR BUBBLES

Even industrial-grade plastics, putties, or resin compounds can't effectively mask all the air bubbles that naturally emerge during the casting of a fake trilobite. Many of the original molds used during this creative process are derived from top-grade "real" specimens. Carefully checking for the telltale signs of surface scarring details is a surefire means of distinguishing a faux fossil from the genuine arthropod article.

EYE DETAIL

Few characteristics of trilobite morphology draw more collector attention and interest than their eyes. That is especially true when it comes to the large compound optics that sit atop the heads of most phacopid species, many of which are noted for their unique preservation and amazing geometric symmetry. When these intricacies are either missing or muted on a specimen that *should* feature such qualities in fine detail, you are probably dealing with a poorly constructed fake.

SURFACE TEXTURE

No matter how good the original mold or how premium the replacement resin or epoxy, it is almost impossible for trilobite manufacturers to reproduce the detailed textural elements of a well-preserved trilobite carapace. Even the most state-of-the-art Russian polymers usually fail to faithfully capture the plethora of pores and pustules that often distinguish an authentic trilobite outer shell.

THE MATRIX

A piece of fossil-bearing sedimentary stone usually displays unmistakable signs of the fracture that initially revealed the enclosed trilobite. If such

A poorly manufactured Chinese "trilobite" concoction.

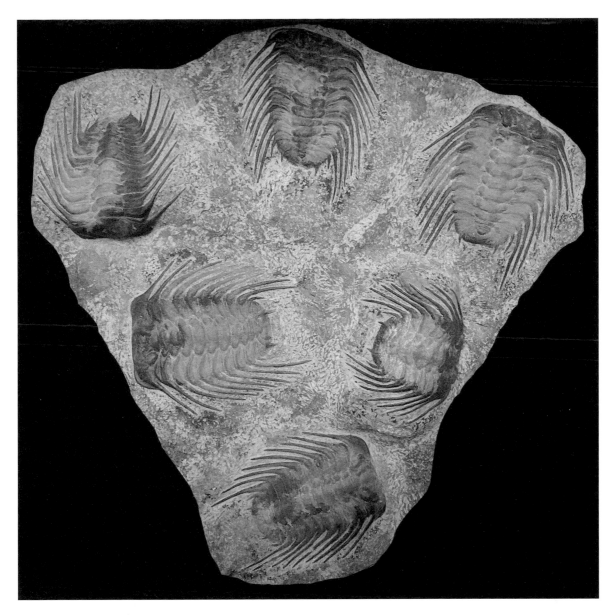

A large plate of 12 cm *Selenopeltis* from Morocco that may (or may not) contain fragments of genuine trilobites was then composited into this "designer" piece.

collateral damage is missing, it could be an indication that a trilobite cast has been artfully yet artificially placed on a separate and "pristine" matrix slab. Although real trilobite specimens may occasionally be remounted on a new stone for cosmetic reasons, an undamaged matrix is usually an unmitigated sign of fossiliferous foul play. Sometimes multiple fake specimens are carefully attached to

the same piece of matrix, creating what some collectors mockingly refer to as a Paleozoic pizza.

USE A BLACKLIGHT

Today small, affordable, surprisingly effective blacklights can be procured online for only a few dollars. When purchasing a trilobite in person, these purple-tinged beams have an unmatched ability to reveal the otherwise "hidden" fill-ins, repairs, and restorations that can occur on even the best specimens. Some modern materials used in manufacturing counterfeit trilobites do not fully fluoresce, yet a blacklight can still expose the area around the fake—especially if the supposed fossil has been crudely attached to a corresponding piece of matrix.

HAND LENS

Whether you're purchasing a perfectly prepared specimen from the most reputable dealer in the world or acquiring a blatant fossil forgery from a backroom scoundrel, using a good hand lens is more than a luxury, it's a necessity. Thus equipped, it is much easier for any trilobite enthusiast to spot the obvious—as well as the clandestine—imperfections that too often connote a fake example.

SEAMS

Some manufactured trilobites emerge from a veritable assembly line of ready-made parts and pieces. Heads, tails, and bodies of various sizes, shapes, and colors can then be artfully crafted together, with the end result occasionally being a reasonably accurate trilobite concoction. Sometimes, however, the seams created by the fabrication process can be spotted with the naked eye or through your trusty hand lens.

THE PRICE

No matter how shrewd a trilobite enthusiast may be, the cold hard fact is that rarely—especially at a large, well-attended fossil and mineral show—are you going to encounter a deal that is too good to be believed. If you see a superficially attractive example of a rare trilobite species being offered at an unusually miniscule price, the unfortunate truth is that it is probably as bogus as a proverbial three-dollar bill.

MORPHOLOGICAL DISPARITY

Sometimes industrious but misguided manufacturers of fake trilobites will haphazardly throw together mismatched parts of various authentic species. In the process of doing so, they unintentionally create chimera-like creatures that would never have seen the filtered light of day in those primal Paleozoic seas. A fundamental knowledge of trilobite anatomy, as well as of the various orders generated by these amazing arthropods during their quarter-billion-year span on planet Earth, should help any enthusiast avoid falling into such a superficially tempting trap.

CROSS SECTION

When all else fails and you are willing to risk it all to determine whether you possess a relatively pristine Paleozoic prize or a mere piece of pretending putty, there is one assertive action you can take to certify your specimen's inherent level of drastic plastic. You can cut the darn thing in half! Such a blatant cross section should forever reveal all you need or want to know about fake trilobites.

27

10 MASS MORTALITY TRILOBITES

T rilobites appear to have been highly communal animals. The fossil evidence shows that they often congregated in closely packed groups, perhaps even traversing the world's seafloors in long, single-file, cephalon-to-pygidia alignments, as one 2016 scientific paper postulates. This behavior was designed to provide safety in numbers, and it subsequently furnished a marked increase in each trilobite's procreative possibilities. In some notable cases, these primal creatures appear to be pervasive within their marine ecosystem. In these repositories, layers of Paleozoic rock have been unearthed that are virtually covered in fossilized trilobite remains. These mass mortality assemblages—such as those dramatically exhibited by the Ordovician asaphid *Anataphrus bromidensis*, where thousands of complete, 6-centimeter-long trilobites have been uncovered fossilized side by side, and one atop the other, in Oklahoma's Bromide Formation—may reflect the end result of an ancient tidal basin draining or evaporating, leaving its inhabitants quite literally high and dry. Other experts state that some trilobites—such as the phacopid *Eldredgeops rana* discovered in the rich Devonian outcrops of upstate New York—may have followed a life cycle that would have drawn members of their species together in prolific numbers at certain times of the year, creating mating or molting conglomerations. The net effect of such behavioral activities may well have left Paleozoic seafloors carpeted with near-complete trilobite carapaces.

Here's a look at 10 trilobites that can be uncovered in mass mortality plates.

REDLICHIA (PTEROREDLICHIA) CHINENSIS WALCOTT, 1905

Late Lower Cambrian; Balang Formation; Western Hunan Province, China; largest 8.4 cm

REDLICHIA CHINENSIS

Among the more common species hailing from the Lower Cambrian strata of China, strikingly similar *Redlichia* trilobites also have been found some 8,000 kilometers away amid the Emu Bay Shales of southern Australia. Multiple plates have been discovered in both localities, although complete examples of the slightly smaller Chinese trilobites (average length 7 centimeters) appear to be more plentiful.

AULACOPLEURA KONINCKI

This diminutive Silurian species from the famed fossil beds of the Czech Republic has long been recognized as being among Europe's most prolific

*AULACOPLEURA
(AULACOPLEURA) KONINCKI
(BARRANDE, 1846)*
**Middle Silurian, Wenlock; Motol Formation;
Loděnice-Barrandovy Jamy, Czech Republic;
largest 1.8 cm**

trilobites. Dating to the mid-nineteenth-century days of the famed Joachim Barrande, multiple plates of these 2-centimeter-long arthropods drawn from the Motol Formation—some featuring dozens of complete *Aulacopleura* examples—have served as the centerpiece for both museum displays and private collections.

BASSEIARGES MELLISHAE

As overwhelming evidence shows, trilobites were social creatures, often living (and subsequently succumbing) in tightly packed groups. This phenomenon is dramatically displayed by the ornate Devonian lichid *Basseiarges* that in recent years has been recovered in multiple plates featuring up to a score of complete fossilized specimens. This unusual species first emerged from the Moroccan dig site of Jorf during the early years of the twenty-first century, and it quickly became a collector's favorite due to its complex morphological profile, unusual preservation, and vivid coloration.

ANATAPHRUS BROMIDENSIS (ESKER, 1964)

Middle Ordovician, Blackriverian Stage; Criner Hills; Carter County, Oklahoma, United States; largest 6.2 cm

AMPYX PRISCUS

Quite a scientific stir was generated in 2016 when it was revealed that the trilobite genus *Ampyx* may have traversed the primeval seafloor in long, single-file formations, with some recovered specimens dramatically displaying this behavior amid their fossilized remains. This evidence of group activity was reported throughout the mainstream media, making members of this species from the Ordovician-age Fezouata Formation of Morocco internationally recognized as true "rock" stars.

GABRICERAURUS DENTATUS

This is one of the most coveted and best-known species hailing from the Ordovician outcrops of North America. *Gabriceraurus dentatus* have occasionally been recovered in mass mortality layers throughout southern Ontario and northern New York State. These attractive 8- to 10-centimeter trilobites have long been favorites of collectors worldwide both for their sleek appearance and their well-preserved exoskeletons.

OLENELLUS CHIEFENSIS

Considered by collectors to be among the keystone Cambrian species to be drawn from the fossiliferous outcrops of the western United States, complete *Olenellus chiefensis* specimens have become almost as rare as unicorn droppings in recent years. Despite their scarcity, on occasion these attractive Pioche Formation bugs have been found in multispecimen plates, examples of which further display both their aesthetic and their academic appeal.

ANATAPHRUS BROMIDENSIS

During the early 1970s, an Ordovician layer dating back some 460 million years was uncovered in Oklahoma that was quite literally covered with the complete remains of the 4- to 6-centimeter long trilobite *Anataphrus bromidensis*, a genus formerly recognized as *Homotelus*. Thousands of these attractively preserved, caramel-toned specimens were found lying one on top of the other, creating a veritable Paleozoic carpet of discarded arthropod anatomy.

ELDREDGEOPS MILLERI

Usually found (and coveted by collectors) as single stand-alone specimens, on occasion multiple examples of the distinctively large-eyed Devonian trilobite *Eldredgeops* have been unearthed. With fossil-yielding outcrops appearing in Ohio and Michigan, midwestern U.S. enthusiasts still regularly march to local quarries—although many are now closed due to safety concerns—in hopes of procuring prime examples of these legendary Paleozoic prizes.

XENASAPHUS DEVEXUS

First uncovered in a now flooded sedimentary exposure located along the Volkhov River in western Russia, *Xenasaphus* plates bearing up to two dozen of these large (10- to 12-centimeter) trilobites were once relatively common sights at fossil shows around the globe. In recent years, however, the Middle Ordovician location from which they had previously been extracted seems to have been worked out, and prime multiple examples are now exceedingly scarce.

PSEUDOGYGITES LATIMARGINATUS

Throughout the later decades of the twentieth century, one of the most frequently visited trilobite-yielding locations in all of North America could be found on the outskirts of Bowmanville, Ontario. In the shales of what is now known as the Blue Mountain Formation, impressively preserved examples of the midsized Ordovician asaphid *Pseudogygites latimarginatus* (with specimens ranging from 3- to 7-centimeters in length) were recovered in large numbers—occasionally in mass mortality plates featuring up to a score of these distinctive white-on-black trilobites.

***GABRICERAURUS DENTATUS* (RAYMOND AND BARTON, 1913)**

Upper Ordovician; Bobcaygeon Formation; Deseronto, Ontario, Canada; largest 8.3 cm

***OLENELLUS CHIEFENSIS* PALMER, 1998**

Lower Cambrian; Pioche Formation; Nevada, United States; largest 5.3 cm

(LEFT) *XENASAPHUS DEVEXUS* (EICHWALD, 1840)

Middle Ordovician, Upper Llanvirnian; Uhaku Regional Stage; Volkhov River, St. Petersburg, Russia; largest 11 cm

(BOTTOM) *PSEUDOGYGITES LATIMARGINATUS* (HALL, 1847)

Upper Ordovician; Blue Mountain (formerly Whitby) Formation; Bowmanville, Ontario, Canada; largest 6.8 cm

28

10 PRECARIOUSLY PREDATED TRILOBITES

Considering the gruesome assortment of lethal-looking bite marks and healed injury scars that frequently adorn their fossilized exoskeletons, it seems safe to surmise that the daily existence of a trilobite was filled with formidable obstacles. The graphic evidence of their hazardous lifestyle offers rather emphatic proof that the primal seas in which these Paleozoic arthropods flourished for nearly 300 million years provided neither a hospitable environment nor a safe path through evolution's treacherous labyrinth. Those ancient oceans were infused with an ever-changing and ever-more-dangerous cast of predatory characters, all seemingly intent on turning the local trilobite population into nothing more than tasty midday morsels. Indeed, in certain locales, such as the famed Upper Cambrian fossil beds of Cranbrook, British Columbia, jagged, monster-made incisions can on occasion be seen adorning the protective calcite shells of the region's abundant *Orygmaspis* and *Labiostria* trilobite specimens. Despite these early efforts to shield themselves, the grim pathologic evidence revealed in sedimentary stones around the globe indicates one thing—even in the subsequent Ordovician, Silurian, and Devonian, trilobites were often the victims of fierce predacious attacks. The discovery of these flagrantly injured fossilized carapaces has provided both scientists and collectors with an additional portal through which to view life in the Paleozoic oceans. After all, it's one thing to examine 500-million-year-old fossils featuring the detailed impressions of ancient life-forms. It's quite another to be able to interpret those impressions in a manner that furnishes greater understanding of what the battle

ELLIPTOCEPHALA SP.
Lower Cambrian; Rosella Formation,
Atan Group; British Columbia, Canada;
9.7 cm

for daily existence may have been like during those long gone yesterdays when trilobites filled the seas.

Here's a look at 10 precariously predated trilobites.

ELLIPTOCEPHALA SP.

Some predatory attacks on trilobites were more savage than others. The example presented here is about as bad as they get—at least considering the slight chance that this Lower Cambrian *Elliptocephala* from the Rosella Formation of Western Canada may have actually survived the initial onslaught! There appears to be some healing along the edges of the bite mark, indicating that a molt possibly occurred following the predation. But perhaps I'm just being overly optimistic.

APATOKEPHALUS INCISUS

On occasion, a trilobite traversing the primal oceans needed to be like an NFL running back—sleek, shifty, and nimble enough to avoid the brunt of serious trouble. That clearly appears to be the case with this 5-centimeter-long *Apatokephalus* from the Ordovician-age Fezouata stratum of Morocco. With a hydrodynamic body design allowing it to move quickly through its marine habitat, it was still not able to avoid what appears to be a predacious attack that saw part of its pygidium neatly clipped.

OLENOIDES SUPERBUS

Due to the often unpredictable nature of their Paleozoic preservation, until the later years of the twentieth century many of the subtle features that distinguish trilobite anatomy—including spines and predation scars—were not easily viewed by either collectors or academics. But as preparation techniques improved by leaps and bounds, such characteristics—as clearly seen on this dramatically devoured Middle Cambrian *Olenoides*—became readily apparent to anyone and everyone.

REDLICHIA REX

No matter how impressive a trilobite may have appeared to be, it seems there was always a bigger, badder bully on the tidal bay block ready to take a predacious bite out of that neighboring arthropod. Evidence shows that the Lower Cambrian species *Redlichia rex*, from Australia's Emu Bay Shale (which grew to lengths exceeding 25 centimeters), was often savagely attacked in its primal marine ecosystem, perhaps by the predacious *Anomalocaris*, a creature also recognized from these layers. Some of the trilobites seem to have survived that initial strike, but they were clearly worse off from the experience.

TRICREPICEPHALUS SP.

Members of the distinctive genus *Tricrepicephalus* rank among the most desired and hard to acquire trilobites in the entire Paleozoic realm. The most recognizable of these uncommon Middle Cambrian trilobites is *T. texanus*, which hails from Utah's renowned Weeks Formation. But many undescribed variants of that attractive species exist in those same Millard County outcrops. This is one of them—with both strange, paddle-like pygidial spines and an intriguing (and apparently healing) bite mark on its right genal spine.

AMECEPHALUS JAMISONI

Scientists have speculated that the predator-prey ratio that existed in the mid-Cambrian oceans may have been somewhat akin to what today exists on the African savannas. Even relatively rare trilobite species such as *Amecephalus jamisoni* may have fallen victim to the fierce creatures that inhabited those primal seas. Their often scarred exoskeletons (here with predation on its left genal spine) bear bold witness to their dangerous undersea status.

ELDREDGEOPS RANA

In some Paleozoic pockets, such as those that appear throughout upstate New York—home to myriad Devonian-age *Eldredgeops* specimens—trilobite fossils are common, as are the predation scars on their dorsal anatomy. But in virtually all of these localities, little or no fossil evidence exists for what may have fed on them. The question then becomes: What predator inflicted these savage wounds? A possible answer is other larger trilobites.

NEVADIA WEEKSI

As soon as there were trilobites in the Lower Cambrian seas, there were predators ready, willing, and able to feast on them. Prime examples of this phenomenon are provided by some of the large (up to 12 centimeter) *Nevadia weeksi* specimens drawn from the 520-million-year-old outcrops of Nevada. Whether or not any of these trilobites were able to survive the blatant attacks perpetrated on them will always be open to speculation and debate. In some cases, however, it appears that the resulting fossil specimen may actually be the munched-upon molt that served as a quick calcium infusion for some voracious undersea creature.

ORYGMASPIS CONTRACTA

For whatever reason (and academia has yet to offer a reasonable explanation), it seems that Cambrian trilobites suffered more than their fair share of predatory attacks. Fewer trilobite fossils from subsequent geologic periods exhibit such graphic evidence of aggression. Perhaps it was the notorious *Anomalocaris*—the terror of the early seas—that took a distinctively shaped bite out of the left-side thoraces of these beautifully preserved, 500-million-year-old *Orygmaspis* and *Labiostria* specimens from the Upper Cambrian rocks of British Columbia.

GABRIELLUS KIERORUM

Even after centuries of exploration and extraction around the globe, new trilobite-bearing locations are still being uncovered on an annual basis. One of the most prominent of these recent Paleozoic discoveries is along the Dease River in western Canada where Lower Cambrian trilobites of notable size, shape, and elegance emerged. A variety of trilobites found in this locality bear small but striking bite marks, including those that frequently adorn the fossilized exoskeleton of the impressive early genus *Gabriellus*, some of which attained dimensions exceeding 8 centimeters in length.

APATOKEPHALUS CF. INCISUS DEAN, 1966

Lower Ordovician, Floian Stage; Upper Fezouata Formation, Outer Feijas Group; Zagora, South Morocco; 5.3 cm

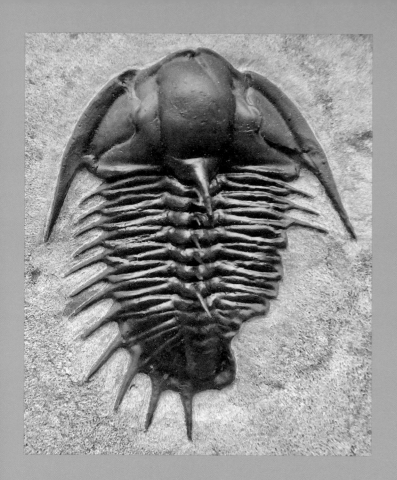

(LEFT) OLENOIDES SUPERBUS (WALCOTT, 1908)

Middle Cambrian; Marjum Formation, House Range; Millard County, Utah, United States; 4.2 cm

(BOTTOM) TRICREPICEPHALUS SP.

Upper Middle Cambrian; Weeks Formation; Millard County, United States; 5.2 cm

(LEFT) *AMECEPHALUS JAMISONI* **(ROBISON AND BABCOCK, 2011)**

Middle Cambrian; Langston Formation; Spence Shale; Miiner's Hollow, Wellesville Mountains; Utah, United States; 3 cm

Note the predation on the left genal spine.

(BOTTOM) *ORYGMASPIS (PARABOLINOIDES) CONTRACTA* **(FREDERICKSON, 1949)**

Upper Cambrian, Furongian; McKay Group, Taenicephalus Zone, Parabolinoides Subzone; Cranbrook, British Columbia, Canada; 6.1 cm

LABIOSTRIA GIBBAE CHATTERTON, 2020

Upper Cambrian, Furongian; McKay Group; Cranbrook, British Columbia, Canada; 4.8 cm

Note the square-shaped bite on the thorax.

29

10 MAGNIFICENT
MOROCCAN TRILOBITES

If you enjoy collecting trilobites, the lure of Morocco is obvious. That colorful North African nation has blossomed into a key cog in the planet's expanding fossil trade over the last four decades. During that time, the emergence of a dizzying variety of trilobite species—ranging in age from Lower Cambrian to Middle Devonian—has helped establish Morocco as one of paleontology's most renowned hotbeds, a place where just about *anything* is available—for a price! When visiting the country's desert-hugging trilobite hub of Alnif, a stroll down that town's notorious street known as Trilobite Alley is like taking a step back in time. The knowledgeable yet shrewd local fossil merchants proudly sit atop the city's paleontological pecking order. It is these men who control the dozen small shops that line Trilobite Alley—many being little more than one room shacks with bare light bulbs illuminating hundreds of specimens sometimes scattered haphazardly across an area-rugged floor. Genera familiar to virtually all trilobite collectors—*Cambropallas*, *Selenopeltis*, *Harpes*, and *Leonaspis*—are all there. They vie for attention alongside more exotic examples—such as *Koneprusia* and *Longianda*—as well as attractive, fresh from the field multiplates, some featuring as many as half-a-dozen complete trilobites, often representing two or more distinct species. On occasion, the authenticity of a few of these ancient arthropod fossils has been rightfully questioned, but the simple fact is that nowhere else on planet Earth are trilobites more diverse and prolific than those found on the sunbaked streets and sand-strewn quarries of Morocco.

Here are 10 particularly notable Moroccan trilobites.

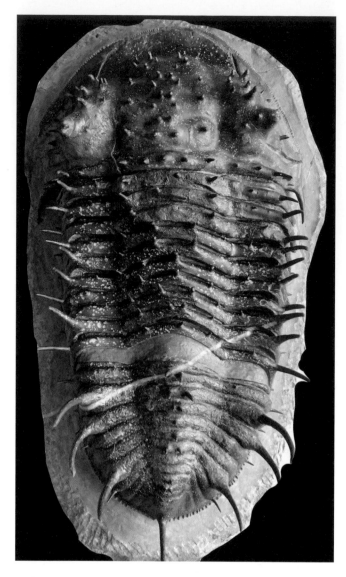

**SCABRELLA (SPINISCABRELLA) STRUVEI
(SCRAUT, 2000)**

**Lower Devonian; Ihandar Formation; Jbel Issimour, Morocco;
24 cm**

Yes, this is a *real* one, though some of the spine tips have been restored.

SCABRELLA STRUVEI

Perhaps the most famous Devonian trilobite in Moroccan fossil lore goes by the nickname "Elvis"—after all, this imposing, spine-encrusted homalonotid is supposedly the King of Trilobites! Dozens of eye-catching—and pricey—examples

of this exotic species have been spotted at fossil shows across the globe. Unfortunately, except for a few scattered pygidia, virtually *all* the Elvis sightings that have taken place in the last three decades have been of fake (or almost totally fake) specimens designed to titillate the trilobite-starved collector's market. Indeed, the consensus opinion is that less than a handful of genuine complete examples of this large (up to 25-centimeter-long) species of *Scabrella* have ever been verified—including the one in the accompanying photograph, although even on that some of the spines have been artfully enhanced.

PLATYPELTOIDES CARMENAE

Members of the *Platypeltoides* genus (which feature seven thoracic segments compared with the eight segments apparent in closely related asaphids) are renowned for growing to prodigious sizes, some more than 40 centimeters in length. Such impressive dimensions mark these trilobites as being among the largest in the world and among the largest yet discovered in Morocco's fertile Paleozoic deposits. *P. carmenae*, from the Lower Fezouata Formation, may not have reached sizes quite as large as those achieved by some strikingly similar but as yet unidentified species hailing from the neighboring Upper Fezouata, but at nearly 20 centimeters this trilobite must have been an impressive sight as it quietly crawled along the Paleozoic seafloor some 430 million years ago.

LONGIANDA TERMIERI

In recent years, more and more unusual and previously unseen and unknown North African trilobite species have begun infiltrating the commercial fossil market. Increasingly exotic examples, such as the impressive *Longianda termieri*—a large, Lower Cambrian species that may represent one of the earliest trilobites emerging from the astoundingly abundant Moroccan Lagerstatte—have been showing up in impressive numbers at leading rock and mineral shows in Tucson, Denver, Tokyo, and Munich. Dozens of dealers, including those from France, Japan, the United States, and England, in addition to the expected Moroccan crews, set up shop at these events, offering a choice supply of spectacular *Longianda* specimens, often at equally spectacular prices.

BRONGNIARTELLA MAROCANA

The first trilobite many novice collectors encounter when visiting a local rock shop or a major fossil and mineral show is a concretion-encased "mud bug" from the Upper Ktaoua Formation of Morocco. These are almost always medium-sized (6- to 8-centimeter) examples of the omnipresent Ordovician species *Flexicalymene ouzregui*, a trilobite that over the last three decades has been uncovered and marketed by the thousands. On rare occasions, something far more unusual is recovered in those same Ktaoua mudstone layers—concretions bearing positive/negative impressions of the rare species *Brogniartella marocana*. Only about a score of these large trilobites have been found, some up to 16 centimeters in length, with many being only partial examples, missing either their heads or tails.

HELIOPELTIS JOHNSONI

Specimens of the delicately spined Devonian trilobite *Heliopeltis johnsoni* are mined from kilometers-long trenches by teams of wandering Moroccan nomads. Over the last 30-plus years, two-meter deep channels have been gouged into the hard limestone overburden with handheld tools, creating a meandering zigzag pattern that runs in accordance with the topographic profile of the surrounding Atlas Mountains. Once a promising rock has been detached, the colorfully clad natives painstakingly break apart every exposed surface, a process that allows them to better view their elusive trilobite targets. Often only a single *Heliopeltis* spine is revealed, but that is enough to alert these experienced and sharp-eyed workers that they have hit upon a promising Paleozoic find.

HUGINARGES ESTEVEI

Hundreds of workers are involved in the day-to-day operations of the active Moroccan trilobite trade—including an intriguing blend of diggers, drivers, preppers, and merchants. Due to the ongoing efforts of this industrious crew, fresh Paleozoic horizons are still being discovered on an annual basis. These new excavations include a series of exciting Devonian quarries that dot the nation's arid interior. Many of these rock-strewn outposts have risen to prominence in the twenty-first century due to the discovery of perfectly preserved examples of previously unknown trilobite species, such as the lichid *Huginarges estevei*. The number and variety of these trilobites have continued to expand in the past few years, with most quickly finding their way into the appreciative hands of the collecting community.

SELENOPELTIS BUCHII

Initially recognized from Ordovician deposits unearthed in central Bohemia (now the Czech Republic) during the middle years of the nineteenth century, closely related species of *Selenopeltis* soon emerged from both the Shropshire localities of the English Midlands and the Valongo Formation outcrops of the Iberian Peninsula. But those time-tested European species positively pale when compared to the dimensions and diversity achieved by many Moroccan members of this spinose genus, some of which feature specimens up to 18 centimeters long. Even in the third decade of the twenty-first century, as new fossiliferous localities are discovered and mined throughout this North African nation, eye-catching examples of previously unseen varieties of *Selenopeltis*—including a large not yet described species from the Lower Ktaoua Formation of the Draa Valley—continue to emerge on the world stage.

FOULONIA SP.

Every year a horde of previously unknown trilobite material—uncovered from increasingly remote Moroccan quarries—appears in the shops in the fossil hub of Alnif. These freshly found specimens can leave even the most seasoned town merchants confused and confounded. Some of these trilobites, including the still scientifically undescribed Ordovician cheirurid *Foulonia*, hail from geological horizons that skirt the steaming desert, and others are culled from Cambrian and Devonian dig sites that sit mere kilometers from downtown Marrakesh. A visitor may ask any of the experienced Alnif dealers about these strange new creatures in either of the city's spoken languages, French or Arabic, but often the best translation of their shrugging response is "your guess is as good as mine." The *Foulonia* genus is notable in fossilized form for possibly revealing the trilobite class's ability to traverse the primal seas in long, single file formations.

PROCERATOCEPHALA SP.

Over the last decade, there has been a steady influx of state-of-the-art pneumatic scribes and air abrasive drills into many of the remote, fossil-filled regions that permeate the North African landscape—along with the corresponding array of sturdy gas-powered generators needed to power them. When placed in the hands of talented Moroccan prep artisans, these tools have helped expedite the painstaking preparation process required to properly showcase delicately spined Ordovician trilobites such as *Proceratocephala*. And to the delight of collectors around the globe, these enhanced prep efforts have markedly improved the overall quality of the resulting specimens. Trilobites hand-prepared in Morocco were once considered crude (if not ruined), but today many of these same specimens represent the pinnacle of the fossil prep art.

HARPES HAMARLAGHDADENSIS

At one time or another even top trilobite collectors, along with a growing number of museum curators and university professors, have admitted to being fooled by first-rate and highly deceptive Moroccan handiwork. Such fraudulent practices have proven to be particularly true with regard to some rare Ordovician genera, including examples of the *Harpes* genus. (The *H. hamarlaghdadensis* pictured here is, of course, the "real thing.") These clever but duplicitous efforts have effectively utilized highly detailed trilobite casts—comprised of everything from cheap automotive plastic, to top-quality resin, to a rough, mud-based cement—placed on real Paleozoic matrix to create a line of faux fossils that have unfortunately infiltrated the international marketplace right alongside the genuine Paleozoic articles.

***PLATYPELTOIDES CARMENAE* (CORBACH ET AL., 2017)**

Lower Ordovician, Upper Tremadocian; Upper part of Lower Fezouata Formation; Guelmim area, Western Anti-Atlas, Morocco; 19.7 cm

***LONGIANDA TERMIERI* HUPÉ, 1953**

Lower Cambrian; Issafen Formation, Zone Vi, Sectigena Zone; Issafen, Morocco; 16.7 cm.

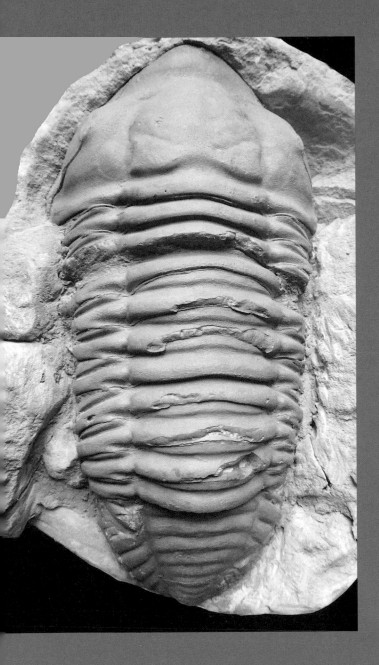

(LEFT) **BRONGNIARTELLA PLATYNOTA
MAROCANA DESTOMBES, 1966**

Upper Ordovician, Lower Ashgillian; Upper Ktaoua Formation;
Bordj, Morocco; 14 cm

(BOTTOM) **HUGINARGES ESTEVEI (CORBACHO
AND KIER, 2013)**

Lower Devonian, Pragian; Ihandar Formation; Jebel Issoumour;
Maider, Morocco; 2.2 cm

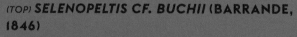

(TOP) *SELENOPELTIS CF. BUCHII* (BARRANDE, 1846)

Upper Ordovician, Caradoc Series; Lower Ktaoua Formation; Blekos (near Erfoud), Morocco; 12.3 cm

(TOP, RIGHT) *PROCERATOCEPHALA SP.*

Upper Ordovician; Lower Ktaoua Formation; Jebel Tijarfaïouine; El Kaid Errami, Morocco; 4.5 cm

(BOTTOM, RIGHT) *HARPES HAMARLAGHDADENSIS* CRÔNIER, OUDOT, KLUG, AND DE BAETS, 2018

Lower Devonian, Upper Emsian; Amerboh Group; Jorf, Tafilalt region, Morocco; 4.4 cm

30

10 PLANET-SPANNING
PARADOXIDES

Soon after the dawn of the Cambrian Period, thousands of distinct trilobite species filled Earth's waters, inhabiting virtually every available maritime niche from surface to seafloor. Never before or since have the oceans been as evolutionarily provocative in their abundance and diversity. This invertebrate bounty has furnished the scientific community with a profusion of key stratigraphic markers that appear in properly aged geologic horizons around the globe. Perhaps nowhere is this paleontological phenomenon more evident than when dealing with the Middle Cambrian trilobite *Paradoxides*. In fact, fossils of closely related—and nearly identical—members of the *Paradoxides* genus have been uncovered in such disparate locations as Newfoundland, Sweden, Massachusetts, Wales, Spain, the Czech Republic, and Morocco. As members of these *Paradoxides* species perished, their calcite-coated debris was buried under millennia of mud and sand deposits. There they remained as the continental masses containing their fossilized residue broke apart and slowly transported these trilobites to their current locations atop the global lithosphere. It is the incredible power behind this tectonic process that explains why a *Paradoxides davidis* specimen uncovered along the rocky cliffs of Wales possesses a nearly identical morphological profile to a *Paradoxides davidis* now found 3,600 kilometers away amid the rugged outcrops of Newfoundland. At a special moment in Paleozoic history some 500 million years ago, these trilobites shared an overlapping habitat off the coast of the early continent of Avalonia. Much the same can be said

***ECCAPARADOXIDES PRADOANUS* (VERNEUIL AND BARRANDE IN PRADO ET AL., 1860)**

Middle Cambrian, Series 3; Upper Caesaraugustian to Lowermost Languedocian; Murero Formation; Sierra del Moncayo, Spain; 6.7 cm

these species all apparently lived in the extended range of the same thriving trilobite community.

Here's a brief overview of 10 planet-spanning *Paradoxides*.

ECCAPARADOXIDES PRADOANUS, SPAIN

This sleek, medium-sized *Paradoxides* species, which can be recovered as positive/negative splits up to 12 centimeters in length, is found in select sedimentary strata throughout the Iberian Peninsula. Often distorted by the planet's tectonic forces, they traditionally appear either elongated or truncated—although such deformation does little to detract from their intrinsic appeal to collectors around the globe.

PARADOXIDES DAVIDIS, NEWFOUNDLAND

These large, impressive specimens—some up to 35 centimeters in length—have been carefully extracted from a small number of outcrops that appear along the rugged Newfoundland coastline. They are nearly identical to *P. davidis* examples uncovered in Wales, now 3,600 kilometers away, a fact that serves as a classic and revelatory example of the continent-shifting powers of plate tectonics.

PARADOXIDES DAVIDIS, WALES

No, this is not a misprint—the trilobite nomenclature here is identical to the previous one. This robust Middle Cambrian species emerges in Wales along a sea-facing cliff wall that dares explorers with both its inaccessibility and its inherent Paleozoic prizes. Despite decades of dedicated exploration, fewer than a score of complete examples have been pulled from the St. David's location, making

for the closely related *Paradoxides gracilis* in the Czech Republic, *Eccaparadoxides pradoanus* in Spain, *Acadoparadoxides harlani* in Massachusetts, and *Acadoparadoxides levisetti* in Morocco. During a crucial period in our planet's distant past,

these trilobites true trophies for any comprehensive collection.

PARADOXIDES GRACILIS, CZECH REPUBLIC

Prior to a global fossil "revolution" during the last decades of the twentieth century, this was unquestionably the most recognized *Paradoxides* species in the world. Examples had been collected and studied since the middle years of the nineteenth century. Often reaching key outlets in western nations even at the height of the restrictive Cold War, these midsized Czech trilobites, which reached lengths up to 18 centimeters, have traditionally ranked among the true bellwethers of any prominent public or private fossil display.

PARADOXIDES PARADOXISSIMUS, SWEDEN

Academically, *Paradoxides paradoxissimus* holds a singular significance for having an "extra" thoracic segment. Most *Paradoxides* feature 20 such segments, but this species has 21. The exact role that such a minimally divergent morphological configuration may have played in this trilobite's evolutionary efforts (other than to serve as a Spinal Tap-esque "one more") is still being debated in scientific circles.

ACADOPARADOXIDES LEVISETTI, MOROCCO

This species is named in honor of Riccardo Levi-Setti, whose late-twentieth-century books on trilobites brought these arthropods to previously unknown levels of public focus and collector acclaim. *Acadoparadoxides levisetti* is most easily distinguished from other Moroccan *Paradoxides*

species by the unusual shape of its relatively diminutive pygidium.

ECCAPARADOXIDES ROUVILLEI, FRANCE

These impressive Middle Cambrian specimens (generally 5 to 9 centimeters in length) are nearly identical in size and morphological configuration to species of *Eccaparadoxides* recovered throughout the Iberian Peninsula. The French examples are perhaps most notable for their wide-ranging colors of preservation that can include tinges of brown, black, and red. Although scarce today, prime *E. rouvillei* examples occasionally can still be found for sale in certain well-appointed European rock shops.

PARADOXIDES SACHERI, CZECH REPUBLIC

Perhaps the rarest of the *Paradoxides* species to emerge from the fossil-laden Barrandian outcrops of the Central Czech Republic, virtually all known "complete" examples of *P. sacheri* are notably missing their free cheeks, indicating that they are molts. Even so, these imposing specimens—some up to 20 centimeters in length—are eagerly sought by both institutions and hobbyists around the globe.

ACADOPARADOXIDES HARLANI, MASSACHUSETTS

Recognized primarily from a single Middle Cambrian outcrop that is now buried beneath a Braintree, Massachusetts, parking lot (along with a handful of complete specimens discovered in similarly aged Newfoundland deposits), remarkably few complete *Acadoparadoxides harlani* have ever been recovered. Many of these—along with a healthy smattering of disarticulated trilobite

***PARADOXIDES DAVIDIS TRAPEZOPYGE* (BERGSTRÖM AND LEVI-SETTI, 1978)**

Middle Cambrian; Manuels River Formation; Manuels River, Newfoundland, Canada; 22.5 cm

fragments, some of which indicate the impressive dimensions these creatures may have once attained—are now housed in the backrooms of Harvard's famed Museum of Comparative Zoology.

ACADOPARADOXIDES BRIAREUS, MOROCCO

Since the early 1980s, *Acadoparadoxides briareus* specimens have been unearthed by the hundreds—if not thousands—in the North African Paleozoic Lagerstatte. Indeed, these uniformly large specimens (some up to 45 centimeters in length) have helped revolutionize the world's perspective on the sizes trilobites could attain. Too often, however, these imposing Middle Cambrian arthropods are recovered as scattered, disassociated bits and pieces that are then haphazardly piecemealed together by eager Moroccan merchants, much to the chagrin of trilobite enthusiasts everywhere.

***PARADOXIDES DAVIDIS* (SALTER, 1863)**

Middle Cambrian; St. David's Group, Davidis Zone; Porth-y-rhaw, St. David's, Pembrokeshire, Wales, United Kingdom; 23.1 cm

Photo courtesy of the Martin Shugar Collection

PARADOXIDES GRACILIS (BOECK, 1827)

Middle Cambrian; Jince Formation; Litavka River Valley, Czech Republic; 13.2 cm

ACADOPARADOXIDES LEVISETTII GEYER AND VINCENT, 2015

Middle Cambrian, Series 3, Agdzian Regional Stage; Jebel Wawrmast Formation, Bèche à Micmacca Member; *Morocconus notabilis* Biozone; Jebel Ougnate (near Tarhoucht), Eastern Anti-Atlas, Morocco; 18.2 cm

ACADOPARADOXIDES HARLANI GREEN, 1834

Middle Cambrian; Manuels River Formation; Manuels River, Newfoundland, Canada; 21.8 cm

31

10 DISTINCTLY DISARTICULATED TRILOBITES

I t is almost impossible to find a place on Earth where the distinctive debris of trilobites cannot be found preserved in various slates, shales, or sandstones. Throughout the globe, paleontologists have named entire geologic horizons for the prolific, frequently disarticulated exoskeletons of these long gone ocean inhabitants. Trilobites evidently grew quite rapidly, and similar to modern arthropods they molted numerous times each year. Thus an overwhelming percentage of their fossilized remains are *not* of the deceased animals themselves but of their cast aside and often fragmentary external armor. In a variety of fossiliferous biozones around the globe, independent heads, tails, and thoracic segments appear to overlay the entire sedimentary strata, with an articulated trilobite rarely in sight. Upon their shells being shed, many of those protective exoskeletons were quickly torn asunder, either by unfavorable undersea conditions or by fellow deep-sea dwellers. When this molting-derived "shrapnel" was combined with the shell residue generated by a legion of deceased trilobite exoskeletons, the resultant trilo-litter may have virtually carpeted certain marine habitats with bits and pieces of arthropod anatomy. Encountering the fossilized evidence of disarticulated trilobite parts is a bane for most serious enthusiasts, but it is a surprisingly satisfying endeavor for many academics who are primarily concerned with examining specific morphological features. In either case, disarticulated parts are an essential and expected aspect of the trilobite collecting experience.

Here's a brief look at 10 distinctly disarticulated trilobites.

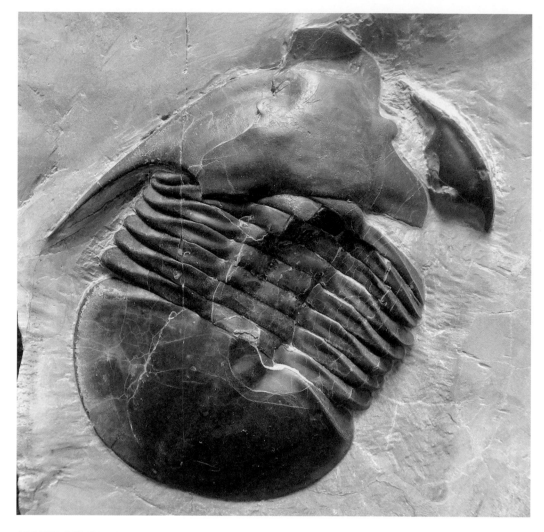

ISOTELOIDES FLEXUS (HINTZE, 1952)

Lower Ordovician; Fillmore Formation, Pogonip Group; Ibex Springs, Utah, United States; 6.3 cm

ISOTELOIDES FLEXUS

Large examples of *Isoteloides flexus* from Utah's 440-million-year-old Ordovician-age Fillmore Formation have been discovered displaying classic evidence of trilobite molting behavior. In these cases, the free cheeks on each side of the animal's cephalon have detached—perhaps along with the entire head itself—allowing the trilobite to pop out of its old exoskeleton and begin the process of generating a new protective shell.

ANOMOCARIOIDES SP.

Any trilobite found on the Great Siberian Plateau is treasured by collectors for both its rarity and its quality of preservation—as well as for reflecting that repository's distant spot on the global lithosphere. Even if a specimen, such as this rare Middle Cambrian *Anomocarioides*, suffers from some degree of dorsal disarticulation, it is embraced by enthusiasts as a unique remnant of Earth's primal past.

DALMANITES DANAE

The Niagaran Limestone Formation of Wisconsin produces some of the most unusual and most coveted trilobites to be found anywhere across the North American continent. Unfortunately, even when a collector is lucky enough to uncover one of the location's Paleozoic treasures, those remnants are often merely disarticulated trilobite heads or tails. On occasion, large examples of *Dalmanites danae*—some up to 11 centimeters in length—have even been uncovered in the midst of molting, creating a somewhat disjointed dorsal appearance.

ACANTHOPYGE HAUERI

Complete specimens of the unusual Devonian lichid *Acanthopyge haueri* first began appearing at fossil shows around the globe in the early 1990s. Unfortunately, as was true for far too many Moroccan trilobites of that era, virtually all of those examples were poorly constructed casts that possessed little or no original trilobite material. Over subsequent decades, perhaps a score of complete authentic specimens have emerged from the fossil-laden deposits of North Africa—although some of those suffer from various states of disarticulation, indicating that they are molts.

TRIMERUS STELMOPHORUS

It is somewhat unusual to encounter a disarticulated *Trimerus* specimen on the open market. There are a number of reasons for this. The first is that the ocean shelf environment in which most of these homalonotids lived some 400 million years ago featured currents that may have torn their molted exoskeletons asunder soon after their deposition. Another reason is that a multitude of complete, ideally aligned specimens have been recovered from the Silurian-age Rochester Shale

ANOMOCARIOIDES SP.
Middle Cambrian, Amgan Stage; Lena River Basin; Siberia, Russia; 3 cm

quarry in upstate New York over the last three decades, so less than perfect examples (such as this rare, seemingly "smiling" example of *T. stelmophorus* from the Lower Devonian strata of Pennsylvania) are simply neither collected nor prepared today.

ALTIOCCULUS HARRISI

Tens of thousands of trilobites have been recovered from the often explored Wheeler Formation

**ACANTHOPYGE HAUERI
(BARRANDE, 1846)**

Middle Devonian; El Otfal Formation;
Mader, Morocco; 8.5 cm

exposures of central Utah. The overwhelming majority of these Middle Cambrian examples are of the ultracommon *Elrathia kingii* and *Asaphiscus wheeleri,* both of which serve as core components of any comprehensive trilobite collection. But hidden away in the same fossil-bearing layers that produce those ubiquitous species are rare examples of *Altiocculus harrisi,* a medium-sized trilobite that on even rarer occasions has been found exhibiting disarticulated (and possibly predated) segments along its thorax and pygidium.

CRYPHAEOIDES ROSTRATUS

Well-preserved examples of the small Devonian trilobite *Cryphaeoides rostratus* (generally under 6 centimeters in length) emerge from their high-altitude Bolivian repositories trapped in hard limestone concretions. These mineralized masses sometimes form around disarticulated trilobite remains, and in that process they create some of the more instantly recognizable fossils found in the Paleozoic realm. Perhaps due to their unusual manner of preservation, a higher percentage of incomplete specimens have been discovered in this location than almost anywhere else on the planet.

GLOSSOPLEURA PACKI

This attractively colored, thin-shelled *Glossopleura* from the Chisholm Shale Formation of eastern Nevada displays intriguing evidence of disarticulation along the left side of its thorax. If you look closely, you can see that this separation was caused when the floating free cheek and genal spine of a *Zacanthoides* trilobite became lodged between the specimen's second and third thoracic pleurae. Although this Middle Cambrian trilobite bears a strong resemblance to *Glossopleura bion* samples found in the neighboring Spence Shale, the rare complete examples discovered in the Chisholm have been identified as *Glossopleura packi*.

FALLOTASPIS BONDONI

This Lower Cambrian species is perhaps most readily distinguished by its uniquely angled genal spines. Disarticulated *Fallotaspis bondoni* head shields are relatively common finds throughout the Zagora region of Morocco where extensive paleontological exploration has taken place over the last decade. But complete specimens are exceedingly rare, and these often serve as the centerpiece of any major Cambrian-centric collection. Here the colorful trilobite has lost its pygidium but maintained the characteristic spines along its axial lobe.

OGYGOPSIS KLOTZI

In certain horizons in British Columbia's famed Burgess Shale quarry, disarticulated trilobite remains are so pervasive that they practically cover the surrounding rock surfaces with the telltale signs of their long-ago lives. There is good reason that so much of this trilobite detritus was generated by *Ogygopsis klotzi*—they were among the largest and most prevalent trilobites found in that renowned Middle Cambrian repository, with some specimens of this ovate arthropod growing to lengths of 10 centimeters or more.

***TRIMERUS STELMOPHORUS N. SP.* BUSH AND SCHWARZ**

Lower Devonian; Shriver Formation; Carbon County, Pennsylvania, United States; 11.7 cm

GLOSSOPLEURA PACKI RESSER, 1935

Middle Cambrian; Chisholm Shale Formation; Pioche, Nevada, United States; 6.4 cm

(LEFT) ***ALTIOCCULUS HARRISI*** (ROBISON, 1971)

Middle Cambrian; Wheeler Formation; House Range, Utah, United States; 5.2 cm

(BOTTOM) ***FALLOTASPIS BONDONI*** (NELTNER AND POCTEY, 1950)

Lower Cambrian, Series 2, Age 3; Issendalenian Regional Stage; Zagora region, Morocco; 5.1 cm

32

10 BEYOND BIG TRILOBITES

The fossil record dramatically demonstrates the astonishing size variance exhibited by Earth's trilobite population. The clear majority of the 25,000 scientifically recognized species that existed during the trilobites' 270-million-year Paleozoic reign were diminutive, only 7 centimeters or less in length. But this information should not be interpreted as indicating that these arthropods didn't routinely grow considerably larger than that. Among the more renowned of these supersized species was *Isotelus maximus*, an asaphid that produced beautifully fossilized specimens exceeding 35 centimeters from tip to tip. During the last two centuries, scores of these smooth-shelled marine behemoths have been drawn from Ordovician-age outcrops throughout the midwestern United States and placed directly (after a little prep work) into the eager hands of collectors around the globe. In addition, during the last 30-plus years, a stunning variety of colossal trilobite species have been found in the prolific Ordovician quarries that border St. Petersburg, Russia. On rare occasions, these formidable examples—including various members of the *Rhinoferus* genus—have exceeded 25 centimeters from head to tail. And since the early 1800s, impressively sized Silurian-age trilobite fossils have been uncovered in New York State, with species such as *Trimerus delphinocephalus* attaining lengths of up to 20 centimeters. In the later years of the twentieth century, the now legendary trilobite beds of northwest Africa began producing Cambrian-age *Paradoxides* specimens of prodigious size, some reaching 45 centimeters in length. At roughly the same time, 60-centimeter-long examples of *Ogyginus*

RHINOFERUS STACYI (SCHMIDT, 1906)
Lower Ordovician; Volkhovian Level; Putilovo Quarry, St. Petersburg region, Russia; 8.7 cm

asaphids belonging to the genus *Rhinoferus*. Some of these specimens can attain lengths in excess of 25 centimeters, and often have pronounced points at the tip of either their cephalon or pygidium, morphological features that continue to perplex both academics and collectors.

ISOTELUS REX

The reigning, defending, undisputed heavyweight champion of the trilobite domain is the Canadian asaphid, *Isotelus rex*. This slightly weathered, 71-centimeter-long monster was found in northern Manitoba in 1998 during surface exploration by a scientific team, and it has since become one of the most photographed and recognized trilobites on the planet. Casts of the specimen are on display in many major institutions, and the original can be viewed in the Manitoba Museum in Winnipeg.

BASILICUS DESTOMBESI

Since emerging as an epicenter of worldwide trilobite activity in the 1980s, Morocco has never ceased to amaze when it comes to the breadth and scope of its Paleozoic holdings. So it should come as little surprise to learn that plates featuring some of the largest asaphid trilobites ever discovered (those of the *Basilicus* genus are up to 50 centimeters in length) emanate from a smattering of recently opened Ordovician deposits that mark the sedimentary outcrops of this North African nation.

NOBILASAPHUS NOBILIS

Since the middle of the nineteenth century, exceptionally large fragments of the Ordovician-aged trilobite genus *Nobilasaphus* have been unearthed in both Morocco and Portugal. Complete specimens (up to 35 centimeters in length) were initially

forteyi were emerging in impressive numbers from the Ordovician outcrops of Portugal's celebrated Valongo Formation. And in 1998, along the shores of Hudson Bay in Manitoba, a 71-centimeter-long specimen of the Ordovician asaphid *Isotelus rex* ("The King of Trilobites") was being collected by Canadian scientists, making it the largest complete, outstretched trilobite yet found.

Here are reports on 10 of the most impressively sized trilobite species ever unearthed.

RHINOFERUS STACYI

The trilobite-bearing deposits of western Russia have continually astounded both collectors and academics with their incredible diversity of Ordovician-age material. Among the more notable examples derived from these fossil-infused layers are a veritable profusion of large, often undescribed

BASILICUS (BASILIELLA) AFF. DESTOMBESI VIDAL, 1998

Lower Ordovician, Middle Arenigian; Upper Fezouata Shales; Draa Valley, Morocco; 38.8 cm

That's a quarter (2.4 cm in diameter) near the right genal spine.

discovered in North Africa in the late 1990s, and less than a decade later crews working in outcrops of the Iberian Peninsula's Valongo Formation found evidence of a remarkably similar species lurking in the fossil-laden shale exposures.

TERATASPIS GRANDIS

For more than a century, tantalizing bits and pieces of the legendary North American lichid *Terataspis grandis* have been found in certain well-defined Devonian outcrops. However, only in the last two decades have a smattering of near-complete examples emerged into the Paleozoic light. A 22-centimeter-long dolomitic *Terataspis* currently resides in Toronto's Royal Ontario Museum, and another articulated specimen is rumored to be a key component of a major Japanese collection.

LICHAS SP.

Deep in the heart of northern Nevada's rugged Cortez Mountains lies the remote Wenban Formation. During the last decade, this Lower Devonian outcrop has begun yielding some of the most remarkable trilobites ever found—with one of the formation's still undescribed lichid species producing complete specimens that exceed 40 centimeters in length. Although still in the first stages of scientific analysis, these dramatic examples of Paleozoic life have made both collector and academic tongues wag from Las Vegas to London.

ISOTELUS LATUS

Since the 1930s, complete specimens of this large, ovate *Isotelus* species (some measuring more than 30 centimeters from tip to tip) have been recognized from Ordovician localities distributed throughout eastern Canada. Indeed, for most of the twentieth century, *Isotelus latus*—along with

the similarly sized *Isotelus maximus* from Ohio—ranked among the more renowned of the planet's oversized trilobites. However, many Canadian quarries have been shutting down in the past few decades, and twenty-first-century discoveries of *I. latus* have become uncommon, with "old collection" examples now eagerly being sought by enthusiasts and institutions around the globe.

OGYGINUS FORTEYI

Canada's *Isotelus rex* enjoys worldwide recognition as the largest trilobite ever found, at a jaw-dropping 71 centimeters. But Portugal's Ordovician-age Valongo Formation has produced a folded example of *Ogyginus forteyi* that is speculated to have been slightly longer—at least before suffering the slings and arrows of postmortem deformation. Named in honor of the British paleontologist Richard Fortey (whose foreword begins this book), it is still hoped that an equally large nondistorted *Ogyginus* specimen will eventually be unearthed at this prolific Paleozoic site.

ASAPHOPSOIDES BREVICA

Even before Chinese fossils became relatively common at international fossil shows in the 1990s, exceptionally large specimens of the unusual dikelocephalid *Asaphopsoides brevica*—some up to 25 centimeters in length—were making their way onto the global market. Import restrictions are now being placed on virtually all fossils emanating from China, so acquisition of a top-quality example of this impressive Ordovician trilobite has become far more difficult.

ZHANGSHANIA TYPICA

Not all oversized trilobites were particularly large. That seemingly oxymoronic statement reflects the

**NOBILIASAPHUS NOBILIS
BARRANDE, 1847**
Upper Middle Ordovician; Upper Llandellian;
Valongo Formation, Portugal; 22 cm

notion that uncharacteristically large examples of a generally midsized species can and do appear in the fossil record on occasion. For instance, most known examples of the Lower Cambrian trilobite *Zhangshania typica* are 3- to 5-centimeters long. At times, however, specimens of this unusual Chinese arthropod have been recovered that reach up to 15 centimeters from head to distinctive split-tail. This disparity may reflect younger and older examples of the same species existing in the same marine niche.

***ISOTELUS LATUS* (RAYMOND, 1913)**

Upper Ordovician; Cobourg Formation; Bowmanville, Ontario, Canada; larger specimen 29.2 cm

ZHANGSHANIA TYPICA LI AND ZHANG IN LI, KANG, AND ZHANG, 1990

Lower Cambrian; Yiliangella-Zhangshania Biozone; Tsanglanpuan Stage, Hongjingshao Formation; Kunming, Yunnan Province, China; 10.2 cm

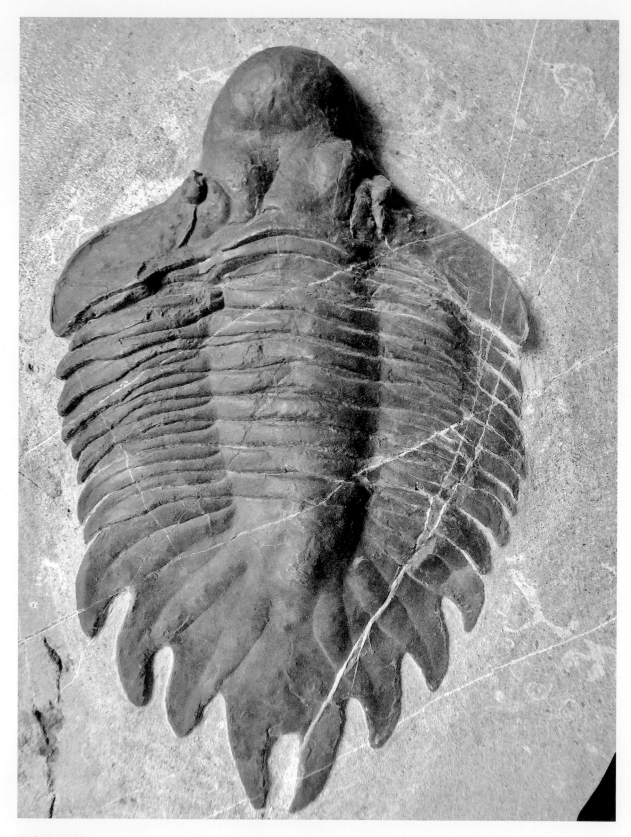

LICHAS SP.

Lower Devonian; Wenban Formation; Cortez Mountains; Nevada, United States; 13.5 cm

33

10 DRAMATICALLY DOLOMITIC TRILOBITES

For many trilobite collectors, the lustrous, calcite-coated outer shells presented by these ancient arthropods represent one of the class's more appealing morphological characteristics. Whether those specimens are well-prepared Devonian phacopids from Ohio or gleaming Ordovician asaphids from Russia, the inherent sheen of their attractive exoskeletons often acts like a proverbial Paleozoic flame, igniting the invertebrate interest of modern-day "moths" from Munich to Milwaukee. However, in certain prime locations across the planet, trilobites emerge from their eons-old sedimentary encasements as little more than dull, dolomite-infused internal molds—specimens totally devoid of those appealingly colorful calcite carapaces. The sites that produce these unusual specimens are ubiquitous, with some of the more renowned dolomitic repositories occurring in Morocco, Scotland, Australia, Wisconsin, and Kazakhstan. Certain enthusiasts—especially those situated in the American Midwest—even seem to favor these dolomitic trilobite specimens above all others. That lure may be based on the uniformity of primal design they present or on the surprising degree of anatomical detail their fossilized remains preserve. The most celebrated source for these U.S. dolomitic trilobites is situated near the town of Grafton, Illinois, where for decades three-dimensional Silurian trilobites of various sizes, shapes, and species have been culled from the Niagaran Limestone outcrops that comprise the surrounding Joliet Formation. The most prominent of these dolomitic bugs includes the 420-million-year-old *Calymene celebra*, which is sometimes found in conjunction with far more unusual trilo-types, including *Cerauromerus hydei*, *Dalmanites platycaudatus*, *Bumastus graftonensis*, and *Ommokris vigilans*.

Here's a look at 10 dramatically dolomitic trilobites.

DALMANITES HALLI

Trilobites of the *Dalmanites* genus rank among the most prevalent on the North American continent with various species appearing in New York, Indiana, Wisconsin, and Pennsylvania. Although many examples sport well-formed outer shells, others—such as *D. halli* from the Silurian layers of Illinois—emerge as mere dolomitic internal molds. This genus also appears in England and Sweden, where the fossilized specimens feature a thick calcite carapace.

CERAUROMERUS HYDEI

This is another of the unusual (and unusually preserved) trilobites that are drawn from the Silurian outcrops of Illinois. Formations including the Edgewood and Joliet produce a variety of three-dimensional fossilized species, including *Cerauromerus hydei*. Until it was recently reclassified, this trilobite was commonly recognized in both the collecting and academic communities as *Cheirurus hydei*.

HADROMEROS WELLERI

Perhaps the rarest Silurian-age cheirurid hailing from the dolomitic outcrops that occur throughout the American Midwest, this diminutive species (usually under 2 centimeters in length) is most notable for the series of short spines emanating along its pygidium. Other recognized examples of the *Hadromeros* genus appear in Scotland and

Quebec, with some of those trilobites attaining lengths of more than 20 centimeters.

DITOMOPYGE KUMPANI

The dolomitic process of fossilization is not limited by either geologic period or geographic region. Even deep in the Carboniferous-age mountains of Kazakhstan, trilobites such as *Ditomopyge* have been found preserved in a tan-toned dolomitic limestone that captures many of the fine details of their dorsal anatomy. Indeed, examples have been uncovered in mass mortality layers surrounded by accompanying flora and fauna. Together these elements provide significant insight into the manner in which these ancient marine creatures lived and died.

TOXOCHASMOPS BISETTI

Emerging from Ordovician deposits located along the rugged coastline of western Scotland, trilobites from the famed Girvan fossil beds—such as *Toxochasmops*—are usually preserved with a colorful mineralized patina ranging in hue from a vivid orange to a rich chocolate-brown. Although never common finds, these specimens have been collected, coveted, and studied for more than 200 years.

OMMOKRIS VIGILANS

No more than a handful of complete examples of this unusual, midsized, midwestern dalmanitid are known to exist, and only a few of those exhibit the full panoply of *Ommokris*'s unusual anatomical features. Occasionally uncovered in close association with other Silurian-age trilobites found throughout Wisconsin and Illinois, the distinctive placement and shape of its stalked eyes mark *Ommokris*

***DALMANITES HALLI* (WELLER, 1907)**

Middle Silurian; Niagaran Limestone, Joliet Formation; Grafton, Illinois, United States; 3.7 cm

**CERAUROMERUS HYDEI
(WELLER, 1907)**

Lower Silurian; Edgewood Formation;
Rochelle, Illinois, United States; 5.2 cm

as one of the more exotic species to appear on the North American fossil landscape.

ODONTOCHILE FORMOSA

Australia's Humevale Siltstone presents one of the most inaccessible fossil-bearing localities on the planet. Located amid the wilds of Victoria, this Silurian outcrop has long been known for occasionally producing dolomitic, internal-mold examples of the rare trilobite genus *Odontochile*. Strikingly similar to a Devonian species found throughout Morocco, these large Aussie dalmanitids are usually poorly preserved and are often unearthed amid other faunal elements, including corals, crinoid stems, and even occasional brachiopod shells. Some experts seem eager to reclassify this into the genus *Zilchovaspis*.

BUMASTUS GRAFTONENSIS

Even in the third decade of the twenty-first century, surprising trilobite discoveries are still being made in locations that have long been known and explored. Amazing specimens of *Bumatus graftonensis* have recently been brought to light in the Silurian-age Hopkinton Formation of Iowa, with some of these trilobites ranking among the largest examples of this genera ever found *anywhere*. One complete dolomitic specimen was measured at more than 13 centimeters in straight-line projection *across* its slightly tucked cephalon!

CALYMENE CELEBRA

Probably the best-known and most readily available dolomitic trilobites in the world are those of the *Calymene* genus that hail from the Niagaran Limestone formations found throughout the midwestern United States. Once so plentiful that they became standard components of virtually every beginner's trilobite collection, in recent years top-quality specimens have become far more difficult to obtain.

ENCRINURUS EGANI

Often uncovered as fragments throughout the Niagaran Limestone, the most recognizable morphological features exhibited by this remarkable encrinurid species are the pair of telescopic eyes that sat perched high atop its cephalon. This was an evolutionary development that allowed the trilobite to remain buried in seafloor ooze while maintaining careful vigil on the surrounding marine community. The fact that dolomitic preservation can effectively fossilize such a seemingly delicate anatomical trait highlights much of its appeal to trilobite collectors worldwide.

(TOP) *HADROMEROS WELLERI*
(RAYMOND, 1916)

Middle Silurian; Niagaran Limestone, Joliet
Formation; Grafton, Illinois, United States;
2 cm

(RIGHT) *DITOMOPYGE KUMPANI* **(V.
N. WEBER, 1933)**

Uppermost Pennsylvanian/Upper
Carboniferous (Stephanian); Ulutau
Mountains; Dzhezkazgan region; Karagandy
Province, Kazakhstan; 2.8 cm

ODONTOCHILE FORMOSA GILL, 1848 AND *PLEURODICTYUM MEGASTOMA* (TABULATE CORAL)

Upper Silurian; Humevale Siltstone, Clonbinane Sandstone Member; Mount Disappointment, Clonbinana District; Victoria, Australia; 13.2 cm

(TOP, LEFT) ***BUMASTUS GRAFTONENSIS* MEEK AND WORTHEN, 1870**

Lower Silurian; Hopkinton Dolomite; Iowa, United States; 13.3 cm across cephalon

(TOP, RIGHT) ***ENCRINURUS EGANI* (MILLER, 1880)**

Middle Silurian; Niagaran Group; Lemont, Illinois, United States; 4.2 cm

(LEFT) ***CALYMENE CELEBRA* (RAYMOND, 1916)**

Middle Silurian; Niagaran Limestone, Joliet Formation; Grafton, Illinois, United States; 4.7 cm

10 EXTRAVAGANTLY EXPENSIVE TRILOBITES

Most hobbyists like to believe that they collect trilobites out of personal curiosity and a quest for knowledge rather than for any potential financial gain. But in recent years collecting these astonishing arthropods has become something of a deep-pocket endeavor, especially for top-tier enthusiasts around the globe. At a certain level, these collections—which may contain literally thousands of specimens—clearly become investments as well as mere recreation. Yes, anyone can still go out in the field and with a bit of luck find a beautifully preserved trilobite in a local quarry, hillside, or streambed. And it's true that a nice, shelf-filling assortment of common species can be sensibly assembled for $500 or less. However, to amass an extensive trilobite collection featuring superlative examples of unusual species gathered from major Paleozoic repositories around the globe, the hobbyist faces a challenging task that can easily cost a small—or not so small—fortune. In addition, during the early decades of the twenty-first century, some trilobite prices have risen to astronomical heights (often in the mid-four-figure range, and occasionally even touching five figures), primarily due to a startling number of previously unseen species emerging from previously unknown localities. Many of these examples, including trilobites found in such distant destinations as Central Siberia, South Australia, and northern Greenland, require excessive time, effort, and economic output simply to reach and recover. All corresponding costs are naturally—and perhaps a tad too easily—transferred to whomever eventually purchases a specimen resulting from these enterprising expeditions. Also, a good number of these exotic examples—including trilobites hailing from established hotbeds such as Morocco and Russia—have benefited from dramatically improved preparation techniques, making them more aesthetically pleasing, scientifically revealing, and unapologetically costly.

Here are reports on 10 extravagantly expensive trilobites.

BUENELLUS HIGGINSI

Along the northern coast of Greenland, outcrops of the Buen Formation have yielded a small but significant Lower Cambrian fauna featuring the fossilized remains of the midsized trilobite *Buenellus higginsi*. Scores of these often poorly preserved ptychopariids have reportedly been uncovered, but a scant few have emerged on the world market because the site is under the strict control of the Danish government. One specimen was supposedly procured by a noted Hollywood celebrity, who forked over the big bucks to secure its acquisition.

APIANURUS SP.

Despite more than a century's worth of callous-causing effort, only one known example of the beautiful, spinose odontopleurid *Apianurus* has been found in the legendary Walcott/Rust Quarry of New York State. That 8-centimeter-long specimen was uncovered early in the twenty-first century. Following more than 100 hours of careful preparation, it was purchased by a private collector who promptly donated it to the Smithsonian Institution. There it remains on display for millions to enjoy.

BOEDASPIS ENSIFER

In the last two decades, a score of complete examples of the large odontopleurid *Boedaspis ensifer* have been uncovered in the fossil-filled Middle Ordovician quarries of western Russia. Although it is a relatively common species—at least when compared to other members of this "most expensive" list—*Boedaspis* (which can reach lengths up to 11 centimeters) continues to amaze collectors around the globe with its morphological complexity and stunning strangeness. For those wondering, it is known that freeing this trilobite's proliferation of spines from its surrounding Ordovician matrix can take up to 80 hours of delicate prep work, adding significantly to its eventual five-figure price on the commercial market.

PSEUDOMERA BARRANDEI

In the rugged desert environs of Nye County, Nevada, lurks the imposing Antelope Valley Limestone. Merely reaching this dry and distant Middle Ordovician outpost has proven to be a deterrent for many fossil-seeking adventurers. But for those who have journeyed to this remote destination, the trilobite rewards—although far from abundant—are certainly noteworthy. Among the most significant is the rare species *Pseudomera barrandei* that, when found complete, represents one of the true treasures of the North American fossil realm.

METOPOLICHAS BREVICEPS

Following decades of determined digging in the famed Waldron Shale quarries that sporadically appear throughout the landscape of southern Indiana, only a single complete specimen of the attractive lichid *Metopolichas breviceps* has been reported. With its mottled shell color and engaging anatomical alignment, this is one of the most coveted of all North American trilobites. It is now housed in a private collection in the United States.

BUENELLUS HIGGINSI (BLAKER, 1988)

Lower Cambrian; Buen Formation; Sirius Passet; Northwest Greenland; 6.2 cm

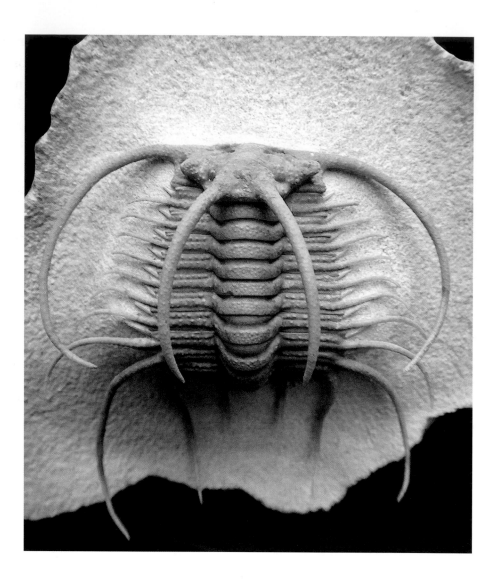

BOEDASPIS ENSIFER (WHITTINGTON AND BOHLIN, 1960)

Lower Ordovician; Volkhovian Level Formation; Putilovo Quarry; St. Petersburg region, Russia; 6.4 cm

LAETHOPRUSIA SP.

Black Cat Mountain in Coal County, Oklahoma, is renowned for producing some of the best-known and best-preserved trilobites ever found in North America. But even among the famous tally of trilobite species that emerge from this often studied rural outcrop, occasional surprises appear, such as the small odontopleurid *Laethoprusia sp.* So far, only one complete example has been uncovered, and that slightly convex 2-centimeter-long specimen now resides in a private collection in Japan.

PLACOPARINA SEDGEWICKII

In the opinion of many notable collectors, these robust, often magnificently preserved, 6- to 10-centimeter-long cheirurids rank among the most desired of all European trilobites. Three distinct species of *Placoparina* hail from the rugged inlier exposures that appear sporadically around the Welsh town of Builth Wells, and these have been eagerly sought by both leading institutions and savvy hobbyists for more than 150 years. Yet despite the long-term fossiliferous focus that has been placed on acquisition of these attractive arthropods, their appearances on the international market have remained infrequent. Only in the last few decades have complete *Placoparina* specimens regularly made their way out of leading European collections.

DEIPHON FORBESI

Deiphon forbesi is one of the smaller members of the *Collector's Guide* "extravagantly expensive" list. Virtually all known examples of this bubble-nosed cheirurid (many collected in the nineteenth century) measure less than 3 centimeters in length. The clear majority of these attractive Silurian specimens—the most famous hailing from the renowned Wren's Nest outcroppings of the English Midlands—are either housed in noted European museums or held as family heirlooms in British cabinets of curiosities.

MEGISTASPIDELLA HEROICA

Among the amazing army of trilobite species that have sprung from the Ordovician-age quarries surrounding St. Petersburg in the last four decades, some deserve special recognition. One of these notable Russian species would certainly be the intriguingly shaped *Megistaspidella heroica*, a trilobite with pronounced points emanating from both its cephalon and pygidium. The specific role these unusual morphological features may have played during this creature's lifetime is still open to speculation and debate, although most contemporary academic wisdom suggests that they may have aided the trilobite in digging through the soft seafloor sediments in which it lived.

STELCKASPIS PERPLEXA

Lake Temiskaming is a large, freshwater body located some 600 kilometers northwest of Montreal. It sits squarely on the provincial border that separates Ontario and Quebec in eastern Canada. Most folks who travel to this thickly forested refuge do so for the swimming, boating, or fishing (over 30 species of fish live in the lake's pristine waters), but others come exclusively in search of trilobites. Abundant outcrops of the Silurian-age Thornloe Formation cover the surrounding landscape, and in recent years the area's rocks have yielded a limited but varied assortment of well-preserved species, including *Stelckaspis perplexa*. Only a handful of articulated examples of this midsized, 2- to 4-centimeter odontopleurid have been unearthed, but ongoing explorations in the vicinity offer the promise of exciting new discoveries.

PSEUDOMERA CF. BARRANDEI (BILLINGS, 1865)

Middle Ordovician; Antelope Valley Limestone; Nye County, Nevada,
United States; 5.9 cm

(TOP) *METOPOLICHAS BREVICEPS* (HALL, 1864)

Middle Silurian; Waldron Shale; Shelby County, Indiana, United States; 6.2 cm

(TOP, RIGHT) *MEGISTASPIDELLA HEROICA* (BOHLIN, 1960)

Middle Ordovician, Lower Llanvirnian; Kunda Regional Stage; Voybokado Quarry; St. Petersburg, Russia; 9.2 cm

(RIGHT) *STELCKASPIS PERPLEXA* (BILLINGS, 1866)

Lower Silurian; Thornloe Formation; Lake Temiskaming mining district; Northern Ontario, Canada; 2.7 cm

PLACOPARINA SEDGEWICKII SEDGEWICKII (McCOY, 1849)

**Middle Ordovician, Llanvirn Series, Llandeilian Stage, teretiusculus Biozone;
Llan-ffawr Mudstone Group; Pencerrig, Builth Inlier, Wales, United Kingdom; 8.1 cm**

Photo courtesy of the Martin Shugar Collection

35

10 SPECTACULARLY SPINED TRILOBITES

During their quarter-billion-year journey through Paleozoic seas, trilobites often came face-to-face with a menacing cast of marine predators and competitors. In response to such dangers, these invertebrates developed an increasingly complex morphological profile that in many ways proved to be as revolutionary as it was evolutionary. To combat such perils, trilobites began growing a series of elaborate and apparently highly effective spines, which in some species, such as the Devonian-age *Drotops armatus*, eventually emanated from virtually every segment of their dorsal anatomies. Exactly when the trilobite class first generated this exotic external ornamentation is a question still open to scientific speculation and debate. It appears that a few early members of the trilobite line, including the 520-million-year-old *Esmeraldina rowei*, did possess a rudimentary form of this adaptation. But by the Middle Cambrian, some 10 million years later, species such as *Doropyge randolphi* and *Olenoides nevadensis* featured an imposing series of spines that both projected in a straight line down their axial lobe and in some cases encircled much of their thorax and pygidium. It has been proposed that these primal spines—constructed from the same calcite-infused materials as their outer shells—may have been somewhat flexible, and perhaps were initially utilized by certain trilobites as a basic form of navigational rudder. Some academics believe that they may also have served an important role in sexual display and courtship. Others have even speculated that the spines may have been attached to nerve-filled sensory organs, providing the trilobite with an ability to perceive disturbances in the water around it. Still others, however, postulate that

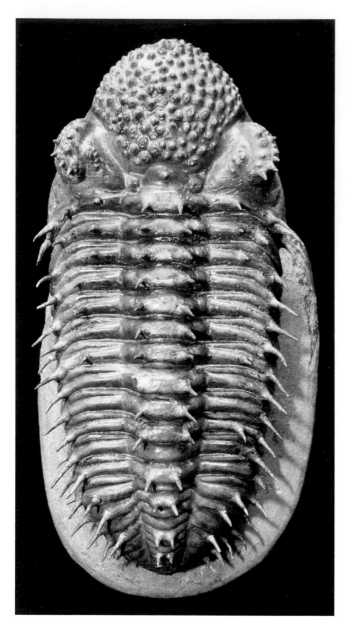

***DROTOPS ARMATUS* FORMA *HOPLITES* STRUVE,
1995**

Middle Devonian, Basal Givetian; Bou Dîb Formation, Lower
Member; Jebel Issoumour; Maïder region (near Alnif), Morocco;
16.7 cm

these primitive spines represented nothing more
than the beginning of an undersea "arms race" that
saw creatures around the globe gearing up for daily
battles of survival.

Here are brief descriptions of 10 spectacularly
spined trilobites.

CERATOLICHAS SP.

For decades, weekend warriors patrolling the
fossil-packed outcrops of eastern Canada would
occasionally stumble on tantalizing, spine-covered
fragments of the rare Devonian genus *Ceratolichas*. Then, in the second decade of this century,
a handful of complete 3- to 5-centimeter examples of this exotic trilobite were uncovered in the
rarely explored (and challenging to pronounce)
Kwatabohegan Formation of Ontario.

COMURA BULTYNCKI

The first *Comura* specimens to emerge from
Morocco in the late 1980s were colloquially known
among collectors as "Mohawk Bugs" because of
the prominent rows of sediment-encased spines
that ran vertically along each trilobite's axial lobe.
Slowly but surely those spines began to be freed
from their surrounding matrix through the efforts
of talented prep artists, and some of the most
astonishing trilobites in the entire fossil record
were subsequently revealed.

ESMERALDINA ROWEI

These sleekly shaped examples from the Lower
Cambrian of California date back some 520 million years. Scientists have speculated that *Esmeraldina* may have been among the first (if not *the*
first) trilobite genus to display pronounced axial
spines. They are often found in a "tucked" position with their tails wrapped under their thorax, a
fossilized stance that serves to further enhance the
presences of those early, jagged-edged spines.

DICRANURUS HAMATUS

This spectacularly spined trilobite is officially
known as *Dicranurus hamatus elegantus*, and for

CERATOLICHAS SP.
Lower-Middle Devonian; Kwatabohegan
Formation; Eastern Ontario, Canada; 3.1 cm

Photo courtesy of the GMR Collection

good reason. With its long, flowing cephalic "horns" and streamlined body design, this rare Devonian species from the fossil-rich outcroppings that dot the landscape of southern Oklahoma is clearly one of the most elegant and renowned members of the entire arthropod phylum. A nearly indistinguishable example of the genus, *Dicranurus monstrosus*, is found some 8,000 kilometers away in similarly aged Moroccan strata, a fact that lends credence to a variety of important scientific concepts, including plate tectonics.

ATRACTOPYGE XIPHERES

From the late 1990s moment when complete examples of the dramatic Upper Ordovician genus *Atractopyge* first appeared at fossil shows around the globe, the entire trilobite community has remained astounded by the graceful configurations displayed by these 450-million-year-old trilobites. With their pronounced pair of pygidial spines, these rank among the more distinctive and easily recognizable species to have emerged from Russia's fossil-filled Viivikonna Formation. Their

ESMERALDINA ROWEI FRITZ, 1995
Lower Cambrian; Waucoban Series; Esmeraldina County. Nevada, United States; 9.1 cm

beautiful caramel color and dramatic three-dimensional preservation have quickly placed this species among the most coveted trilobites in the world.

APIANURUS SP.

It took countless hours of painstakingly slow and detailed work for professional trilobite preparators to free the only known complete specimen of *Apianurus* from its thick, limestone encasement.

Hailing from the famed Walcott/Rust Quarry in upstate New York, this 8-centimeter-long odontopleurid was recognized only from scattered bits when that Ordovician quarry first began operations in the late nineteenth century. A twenty-first-century discovery finally managed to put all those spinose pieces into their proper Paleozoic perspective. This unique example currently resides in the Smithsonian where it is on prominent display in the museum's new fossil hall.

RADIASPIS SP.

Some trilobite specimens can be fully appreciated only after skilled preparators have completed the detailed rock-removing magic that is their stock in trade. Atop the ever-lengthening compendium of prep-enhanced trilobites would certainly be the strange, incredibly spinose, Middle Devonian genus *Radiaspis*—one of the more dramatic (and difficult to prepare) examples ever to emerge from the fossil-rich Moroccan Lagerstatte. Less than a score of complete specimens of this 5- to 8-centimeter-long odontopleurid have been successfully exhumed so far.

OLENOIDES NEVADENSIS

When the first complete *Olenoides* were uncovered in the Middle Cambrian layers of Utah during the middle years of the twentieth century, they were almost always found as positive/negative splits. It wasn't until some three decades later that certain collectors became willing to risk the welfare of their specimen's "negative" half by subjecting it to modern preparation techniques. If they were lucky, the resulting reassembled trilobite displayed a prominent row of axial spines, all of which had been previously buried in the split's reverse side.

SELENOPELTIS SP.

If *Dicranurus* is the most acclaimed of all Moroccan spiny trilobites (their elegant conformation having been thoroughly addressed in other sections of this book), then genera such as *Selenopeltis*, *Ceratarges*, and *Ceratonurus* all rank a very close second. With their long, smoothly curving spines, it has been speculated that various species of the Ordovician-age *Selenopeltis* may have lived at least some of their lives gently floating in strong marine currents. At other times, they utilized their wide morphological form to comfortably reside along the seafloor.

HOPLOLICHAS PLAUTINI

With a pincushion-like profusion of spines emanating from its glabella, *Hoplolichas* was one trilobite apparently well suited to face any maritime challenge that may have meandered its way. Unfortunately, much to the chagrin of collectors across the globe, soon after this bizarre species started showing up at fossil shows from Tucson to Tokyo at the turn of the century, news emerged that many of those sharply pointed barbs had been carefully and artificially "enhanced" by Russian prep artisans. The slightly less spinose example presented here shows the way *H. plautini* most likely appeared during its time in the Paleozoic seas.

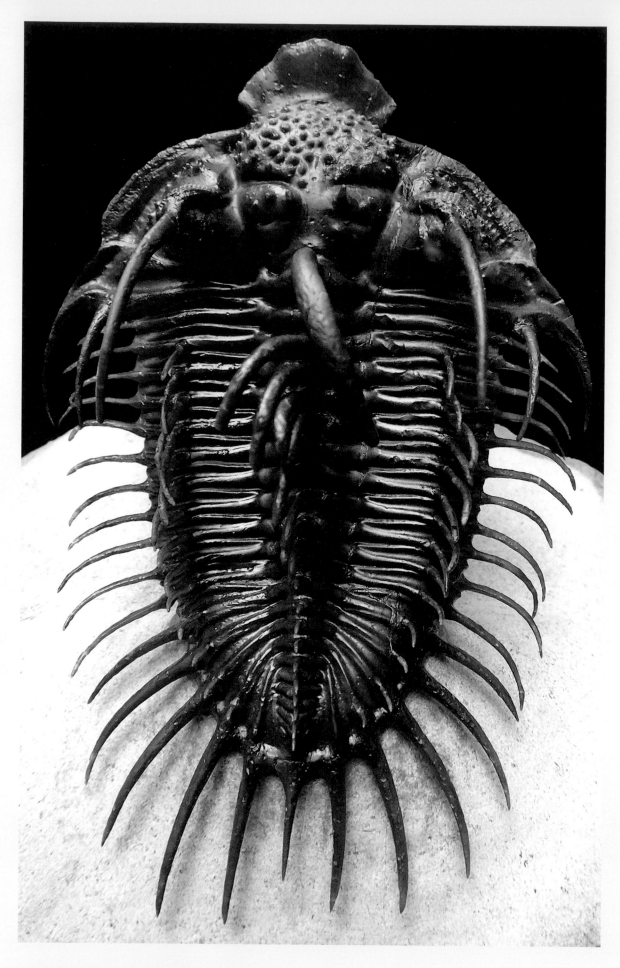

COMURA BULTYNCKI MORZADEC, 2001

Lower Devonian, Upper Emsian; El Otfal Formation; Jebel Oufatène, Maider, Morocco; 6.8 cm

***ATRACTOPYGE XIPHERES* (ÖPIK, 1925)**

Upper Ordovician, Caradocian; Kukruse Regional Stage; Viivikonna Formation; Alekseevka Quarry; St. Petersburg, Russia; 3 cm

(LEFT) *RADIASPIS SP.*

Middle Devonian, Eifelian; El Otfal Formation; Jebel Ofaténe, Morocco; 6.6 cm

(BOTTOM) *SELENOPELTIS SP.*

Upper Ordovician, Caradoc Series; Lower Ktaoua Formation; Tazzarine, South Morocco; 13.2 cm

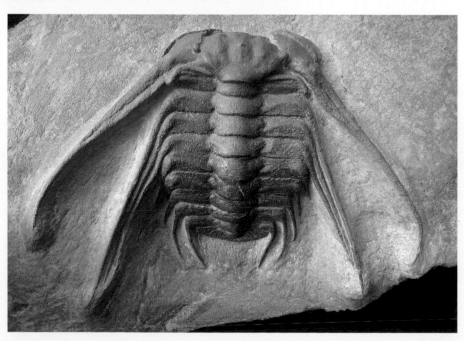

HOPLOLICHAS PLAUTINI (SCHMIDT, 1885)

Middle Ordovician, Upper Llanvirnian; Aseri Regional Stage; Duboviki Formation; Vilpovitsky Quarry, St. Petersburg region, Russia; 6.8 cm

***MEGISTASPIDELLA OBTUSA* (SCHMIDT, 1906)**

Middle Ordovician, Lower Llanvirnian, Kunda Regional Stage; Sillaoru Formation; St. Petersburg region, Russia; 8.2 cm

36

10 TALES TOLD BY TRILOBITES

I f you look closely, every trilobite fossil tells a unique tale. Whether it's a 500-million-year-old *Glossopleura* bearing a potentially fatal bite mark or a 450-million-year-old *Cybantyx* forever frozen in midmolt, each example of these primeval life-forms possesses the capacity to provide a captured-in-time glimpse of a long gone undersea drama. With a trilobite in hand and a healthy imagination at play, each of us can partake in a private sojourn back to a time when trilobites filled the planet's Paleozoic oceans. Indeed, as modern preparation techniques have become increasingly efficient, more explicit details of trilobite anatomy have become known. Corresponding theories concerning the lifestyles of these intriguing invertebrates have also emerged, with each discovery adding immeasurably to the stories every trilobite fossil can reveal. Morphological features such as 3-centimeter-long dorsal spines and wraparound compound eyes provide bold evidence of trilobite defensive postures and unprecedented evolutionary advances. Recently detected signs of fossilized eggs, gills, and gut tracts have afforded both scientists and collectors with previously unavailable and unimagined views into the intriguing domain inhabited by these primitive arthropods. When considered collectively, each of the 25,000 trilobite species that arose during their 270-million-year trek through the Paleozoic reveals an amazing and unique story. Every trilobite fossil, from the most common *Elrathia* to the most exotic *Probolichas*, serves as a vital cog in the complex "machine" of evolution.

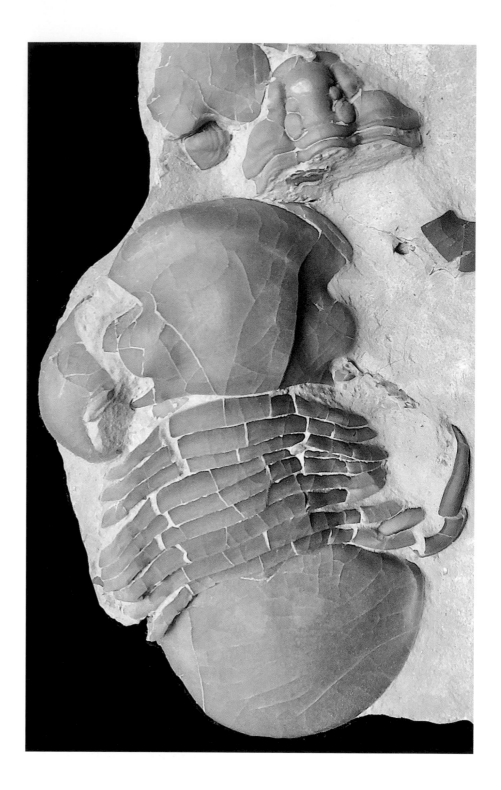

CYBANTYX SP.

Middle Silurian, Sheinwoodian; Osgood
Formation; Ripley County, Indiana,
United States; large molt 8.2 cm

Each tale they tell sheds new light on the darkest depths of Deep Time—the time when trilobites ruled the seas.

Let's take a look at 10 tales told by trilobites.

CYBANTYX SP.

Few more basic trilobite tales can be told than by a specimen captured in midmolt. In fact, fossils of molted trilobite exoskeletons are far more common finds than those representing their perfectly preserved arthropod brethren. Usually, however, such cast aside carapaces—as seen on this *Cybantyx* specimen from the Silurian outcrops of Indiana—are rapidly torn asunder, either by scavenging predators looking for an extra dose of calcite in their diets or by strong undersea currents. And, yes, that is a disarticulated *Calymene* cephalon to the right of the *Cybantyx*.

AMPYX REYESI

Take a good look at the fossil in the photograph of this species. At first it appears to feature a well-preserved example of the unusual Argentinian trinucleid *Ampyx reyesi*—and that's exactly what it does. On closer inspection, however, you will note that this Ordovician trilobite is residing *inside* of a 15-centimeter-long straight shelled cephalopod, where it may have retreated for safety from myriad marine menaces looking to turn it into sushi-in-a-shell.

ARCTINURUS BOLTONI AND BRITTLE STAR

Possible interactions between trilobites and other faunal elements in their primal realm always make for interesting—and often controversial—conversation. Was this imposing specimen of *Arctinurus boltoni* from the Rochester Shale quarry of New York possibly feeding on the *Protaster* brittle star found fossilized in such close association? With a cursory glance, such an argument can certainly be made. But on more detailed inspection, it appears that the trilobite is actually lying a few millimeters above its prospective prey, indicating that these creatures were buried in the mid-Silurian primal ooze at slightly different times. Whether that dramatic difference consisted of hours, days, or weeks we will never know.

TRIARTHRUS EATONI

Few twenty-first century discoveries rocked the paleontological world with more unexpected force than the 2016 revelation that trilobite eggs had been found in Ordovician-age *Triarthrus eatoni* specimens hailing from the famed Beecher's Trilobite Bed of upstate New York. Although long sought, trilobite eggs had previously not been documented in the fossil record. Some of these examples—beautifully preserved in a golden pyrite—feature tiny "specs" that may represent multicell zygotes.

ELDREDGEOPS MILLERI AND CRINOID

On rare occasions, the inherent beauty of a properly prepared fossil manages to supersede its implied scientific gravitas. Such is the case with this Devonian seafloor scenario featuring a large *Eldredgeops milleri* trilobite drawn from the famed fossil quarry in Sylvania, Ohio. By the way, despite their flower-like appearance, crinoids—such as *Arthroacantha carpenteri*—were in fact echinoderms, closely related to starfish and sea urchins.

CALYMENE AND CRINOIDS

Intentional interaction often occurred between many of the organisms that lived along the primal seafloor. On occasion, this synergy was captured in fossilized form, as can be seen in certain *Calymene* specimens recovered from the renowned Silurian outcrops of eastern Quebec. These moment-in-time displays present a detailed depiction of what daily marine life may have been like in the Paleozoic Era, when trilobites apparently utilized the surrounding fauna to their protective advantage.

***AMPYX REYESI* BENEDETTO AND MALANCA, 1975**

Lower Ordovician, Arenig; Acoite Formation, Eastern Cordillera; Northwestern Argentina; 3.4 cm

Trilobite appears inside a 15-cm-long straight shelled cephalopod.

CERAURUS PLEUREXANTHEMUS AND *FLEXICALYMENE SENARIA*

When Ordovician-age sedimentary layers covered in multiple trilobite species were first discovered in Ontario, Canada, in the 1970s, scientists were amazed by the 450-million-year-old stories revealed by these frenetic Paleozoic plates. The stratum bearing these mass assemblages—which produced rock slabs each featuring up to 100 examples of the relatively abundant genera *Ceraurus* and *Flexicalymene*—was quickly and thoroughly collected at the time, and no similar trilobite treasure troves have subsequently emerged from the area's fossil-bearing formations.

BUMASTUSTOIDES MILLERI AND NAUTILOID

When this Ordovician-age nautiloid was first recovered in Quebec, Canada, the lucky collector had little idea that a complete, beautifully preserved *Bumastoides milleri* trilobite was lurking inside the shell's front chamber. This unique relationship was only made evident after some delicate preparation work revealed the fossil within a fossil. Apparently, trilobites frequently used the surrounding fauna (see *Ampyx* and *Neoproetus*) as protective shields to ward off both enemies and the brunt of violent undersea storms.

NEOPROETUS INDICUS AND GONIATITE

As trilobites approached the end of their Paleozoic passage in the Permian, they were continually challenged to seek safe haven from an ever-expanding field of predators. Some, such as this rare *Neoproetus indicus*, apparently sought protection within the convoluted swirls of a neighboring *Metalegoceras* goniatite. Throughout the world—as well as throughout the various time periods in which they lived—trilobite fossils have been uncovered displaying their ability to find shelter in the surrounding marine ecosystem, a factor that undoubtedly aided them in their long-term survival. These temporary refuges included everything from coral colonies and coiled shells to swaying fields of crinoids.

GLOSSOPLEURA GIGANTEA

The Paleozoic seas were a savage place. From the dawn of the Cambrian right through to the end of the Permian—more than a quarter-of-a-billion years later—trilobites constantly battled for survival against an ever-changing and dangerous cast of marine predators. The carapace of some large *Glossopleura* trilobites from Utah bear witness to savage attacks perpetrated on them by unknown undersea monsters. These examples provide us with a compelling story of life (and possible death) in the Middle Cambrian oceans.

ARCTINURUS BOLTONI (BIGSBY, 1825) AND ***PROTASTER*** BRITTLE STAR

Lower Silurian; Rochester Shale Formation, Middleport Quarry; Middleport, New York, United States; trilobite 11.5 cm

TRIARTHRUS EATONI (HALL, 1838)

Upper Ordovician; Lorraine Shale, Martin Quarry; Beecher's Trilobite Bed, Oneida County; New York, United States; largest trilobite 1.2 cm

Photo courtesy of Markus Martin

(TOP, LEFT) ***ELDREDGEOPS MILLERI*** **(STEWART, 1927) AND** ***ARTHROACANTHA CARPENTERI*** **(CRINOIDS)**

Middle Devonian, Givetian; Silica Shale Formation; Sylvania, Lucas County, Ohio, Unite[d] States; trilobite 7.3 cm

(TOP, RIGHT) ***BUMASTOIDES CF MILLERI*** **(BILLINGS, 1859) AND** ***PLECTOCERAS JASONI*** **NAUTILOID**

Upper Ordovician; Trenton Group, Neuville Formation, Grondines Member; Quebec City area, Quebec, Canada; trilobite 3 cm

(LEFT) ***NEOPROETUS INDICUS*** **TESCH[,] 1923 AND** ***METALEGOCERAS SUNDAICUM*** **GONIATITE**

Lower Permian; Maubisse Formation; West Timor, Indonesia; Goniatite diameter 25.6 cm

37

10 WAYS TO VALUE YOUR TRILOBITES

At some point in their collecting career, every trilobite enthusiast must face the question of how to best value a specimen that is a key part of their Paleozoic holdings. After all, a fossil isn't a commodity like oil, gold, or corn to which a preordained international value can be affixed on a daily basis. Natural history items fail to enjoy the economic stability even allotted to works of art, a market in which prices for individual paintings or sculptures can notoriously rise and fall on the whims of a fickle buying public. Quite simply, where trilobites are concerned, values can be unabashedly arbitrary. Their prices are occasionally based on little more than supply and demand. Here the timeworn axiom rings true—beauty, and any corresponding monetary worth, is very much in the eye of the beholder. For example, one of the hundreds of available *Asaphiscus wheeleri* specimens from Utah may be priced at $80, whereas the first example of a previously unknown ondontopleurid species from Russia holds the possibility of being priced at thousands more, and how much more is truly anyone's guess. Of course, numerous commercial websites, international auction houses, and accredited fossil dealers are only too willing to assign a lofty value to the *au courant* trilobite treasure in their possession. Too frequently, collectors are left wondering if the specimen they are being offered has been priced fairly or whether they are just being treated as the latest sucker to walk down the primrose Paleozoic path.

Here's a list of 10 ways to value your trilobites.

SIZE

Yes, size matters. Make no mistake about it, in the realm of trilobites, quite often bigger is better. Although some of the rarest trilobite specimens—whether they are Cambrian, Ordovician, or Devonian—may only measure a few centimeters in length, examples that command the greatest commercial value and generate the most collector interest are usually those that reach dimensions of 6, 8, or 10 centimeters or more. Indeed, it is usually a heady combination of factors—with size perhaps being paramount among these—that can best dictate a trilobite's true monetary value.

ATRACTOPYGE SP.

Middle Ordovician, Lower Llanvirn Series; Shelve Formation, Hope Shale Member; Whitsburn, Shropshire, England; 3.1 cm

RARITY

Every trilobite collection is comprised of the expected assortment of common and semicommon species: *Elrathia kingii, Ellipsocephalus hoffi, Asaphus expansus*, etcetera, etcetera. But it is the choices a hobbyist considers after the acquisition of such anticipated arthropod fare that eventually serves to separate a respectable trilobite grouping from a world-class assemblage. Some specimens are one-of-a-kind treasures; others are rare, beautiful, and pricey examples of trilobite morphological diversity. However, unless you are either incredibly lucky in your field pursuits or know a plethora of top-flight suppliers, quarry operators, or fellow fossil fanatics, obtaining examples of these super rare, invariably valuable bugs can quickly become the often unrequited obsession of any serious trilobite collector.

QUALITY

The quality of a trilobite usually goes a long way toward determining its eventual value. Even complete examples of the rarest species may be quickly dismissed by most hobbyists if they are little more than bruised and battered Paleozoic remnants that have suffered the misfortune of either poor collection practices in the field or less-than-successful subsequent prep work. But modern preparation techniques have helped transform many of these once scarce trilobites into true works of natural art, half-a-billion-year-old relics of unequaled beauty and commercial appeal. Quite simply, the better the quality of your specimen, the higher its potential value.

LOCATION

The location where a trilobite is originally discovered can play a major role in dictating its eventual price and its subsequent value. In recent years, collectors have grown accustomed to seeing an impressive array of new species from Russia and Morocco whenever they travel to an area fossil show or venture onto the internet. However, there is an unusual by-product associated with this level of familiarity that has served to diminish the perceived value of some of these trilobites. Let's face it, if you pass on one magnificently prepared Moroccan arthropod, there is almost always another waiting in the wings. In recent years, equally appealing—and often more valuable—trilobites have begun emerging from such distant destinations as central Siberia, southern China, and northern Greenland. The exotic nature of these remote ports of call directly affects the value of all specimens resulting from the enterprising expeditions needed to acquire them.

SCIENTIFIC GRAVITAS

A trilobite's potential academic importance may not sit atop many collectors' lists of most desirable qualities. But there's no denying that a specimen that has generated significant scientific excitement—such as a *Triarthrus eatoni* bearing eggs or a *Megistaspis hammondi* showcasing antennae and previously unknown soft-body parts—adds immeasurably to both its level of collector interest and its prospective commercial value. In the last few decades, many of the trilobite species at the center of acclaimed academic analysis were first brought to public attention through their discovery by amateur enthusiasts or their subsequent acquisition by dedicated collectors.

PREPARATION

The twenty-first century has proven to be a veritable Paleozoic "gold mine" for trilobite collectors. A good number of the most exotic arthropod

**DAMESELLA PARONAI
(AIRAGHI, 1902)**

Middle Cambrian; Zhangxia
Formation; Shandong Province, China;
10.2 cm

examples ever viewed by the eyes of *Homo sapiens*—especially those hailing from primal hotbeds such as Morocco and Russia—have benefited from dramatically improved preparation techniques, making them more aesthetically pleasing, scientifically revealing, and unapologetically expensive. Even a relatively common trilobite species can see its value doubled or even tripled when subjected to the state-of-the-art skills of a talented prep master, one who can reveal previously obscure surface ornamentation or properly showcase delicate spines or eye facets.

SPECIES

An undeniable truth is that a trilobite's taxonomic order often serves as a key factor in determining its eventual value. Due to their comparative

rarity and corresponding desirability for collectors around the world, lichids such as *Metopolichas hubeneri* are almost invariably more valued than more ordinary phacopids such as *Eldredgeops milleri*. As new species continue to be unearthed in distant Paleozoic corners across the face of the planet, supply and demand must be added to this rather whimsical economic equation. Sometimes even freshly unearthed asaphids or trinucleids (usually among the more common and affordable trilo-types) can ring up a significant score in the ever-changing trilobite value game.

AVAILABILITY

One of the primary reasons trilobite prices have risen so dramatically in recent years can be directly attributed to their increased availability on the global market. Any specimen, at any time, can now be placed on an international forum such as eBay, and prospective buyers from Rome to Rio can then bid against one another to their heart's content. The net result of such auction action has prompted more trilobite material than ever to appear on these outlets, conversely making the prices for particularly desirable specimens spiral continually upward. A nice, cabinet-filling collection of relatively common species can be assembled for a few hundred dollars. But to construct a comprehensive trilobite display featuring top-rate examples of unusual species gathered from major Paleozoic outcrops around the globe, one faces a challenging task that can easily run an enthusiast well over $100,000. It can only be hoped that such a significant initial outlay will translate into a long-term escalation in values.

HISTORY

For some collectors, a trilobite species with "history"—whether it's an example of the famed Dudley Locust, *Calymene blumenbachii*, or the first trilobite presented in scientific literature, *Ogygiocarella debuchii*—increases both its collectability quotient and its inherent value. Certain trilobites have enjoyed long and storied histories; in fact,

hand-drilled trilobites apparently worn as amulets have been uncovered in European archaeological sites dating back more than 15,000 years. As early as the tenth century, Chinese homes commonly displayed trilobite specimens as honored works of natural art. Three hundred years ago, Native American tribes were known to carry small *Elrathia kingii* trilobite fossils in their medicine pouches, regarding them as sacred talismans designed to supernaturally ward off the powers of their enemies.

COLLECTOR DEMAND

With the advent of internet sales in the early 1990s, trilobites that were once relegated to small, table-top displays at local rock and mineral shows were suddenly transformed into natural history collectibles that could be successfully marketed to the entire world. Whether one was ensconced on a fireside ottoman in a London flat or perched atop a poolside chaise lounge in Los Angeles, each and every interested party could then "battle" against one another for the trilobite of their dreams—with the resulting costs of such online activities (as well as any subsequent values) often rising to unexpected and previously unprecedented levels.

ASAPHUS SP.

Lower Ordovician, Tremadoc; Hunan Province, China; 16.6 cm

ELLIPTOCEPHALA SP.

Lower Cambrian, Series 2; Poleta Formation; Montezuma Range; Esmeralda County; Goldfield, Nevada, United States; 5.2 cm

(OPPOSITE PAGE) **METOPOLICHAS SP.**

Upper Ordovician, Caradoc—Early Ashgill; Pupiao Formation; Shidian, Baoshan, Yunnan Province, China; 6.1 cm

(TOP) **BRISTOLIA MOHAVENSIS (CRICKMAY IN HAZZARD, 1933)**

Lower Cambrian; Latham Shale, Marble Mountains; San Bernardino County, California, United States; 6.2 cm

(BOTTOM, RIGHT) **DISTRYAX SP.**

Lower Silurian; Thornloe Formation; Lake Temiskaming area; Northern Ontario, Canada; 5.1 cm

Photo courtesy of the M. Haensel Collection

FAILLEANA INDETERMINATA (WALCOTT, 1877)

Middle Ordovician; Platteville Formation; Prosser, Minnesota, United States; 6.1 cm

38

10 DOPPELGANGER TRILOBITES

Some trilobites seem to possess their own distinct—and at times distantly located—doppelganger. Perhaps the most renowned example of this primeval confluence occurs with the exotic Devonian genus *Dicranurus*. This trilobite, with its long, ramlike cephalic "horns," ranks among the most famous Paleozoic organisms to be recovered from the fossil-laden deposits of Morocco where it is recognized as *Dicranurus monstrosus*. Due to the planet-shifting forces supplied by plate tectonics, a nearly identical example of the genus, *Dicranurus hamatus elegantus*, is found amid similarly aged rocks in Oklahoma, now more than 8,000 kilometers away. In addition, species of the unusual dikelocephalid *Hungioides* have been uncovered in the Ordovician rocks of China, and another closely related example of that rare genus has emerged from the charcoal-hued slate blocks located in Portugal's Valongo Formation. The pervasive Silurian homalonotid *Trimerus delphinocephalus* has been discovered throughout 430-million-year-old layers in upstate New York, and mirror-image species can be noted in both the distinctive limestone horizons of Dudley, England, and the fossiliferous outcrops of central Indiana. As documented elsewhere in this book, nearly identical examples of the Middle Cambrian species *Paradoxides davidis* have been unearthed in both the mudstone repositories of Newfoundland and amid the rugged cliffs of Wales. Even earlier in the trilobites' crawl through evolutionary time, various *Olenellus* species inhabited a relatively finite swath of the Lower Cambrian sea. However, after tens of millions of years of centimeter-by-centimeter movement trapped in the planet's static

sedimentary strata, amazingly similar examples of their distinctive fossilized forms can now be found everywhere from British Columbia, to California, to Pennsylvania, to Scotland.

Here's a look at some of planet Earth's most notable trilobite doppelgangers.

DICRANURUS

With its elongated morphological profile and the sweeping pair of ramlike horns atop its head, *Dicranurus* is one of the most instantly recognizable trilobites in the world. And that's true whether the specimen in question is a *Dicranurus monstrosus* from Morocco or the nearly identical *Dicranurus hamatus elegantus* from Oklahoma, now located nearly a quarter of the planet's distance away.

ECCOPTOCHILE

Eccoptochile represents one of the keystone Ordovician genera found in many comprehensive European trilobite collections. Although rarely discovered complete, the more prominent species of this graceful cheirurid are *Eccoptochile clavigera* from the Czech Republic and *E. mariana* from France, but nearly indistinguishable examples have been unearthed everywhere from England to North Africa.

DIPLEURA

Strikingly similar examples of this Devonian genera can be found in both the limestone quarries

(OPPOSITE PAGE) **DICRANURUS HAMATUS ELEGANTUS (CAMPBELL, 1977)**
Lower Devonian; Haragan Formation; Coal County, Oklahoma, United States; 10.1 cm

of Pennsylvania and the mudstone concretions of Bolivia. The North American specimens are considerably larger (reaching lengths up to 22 centimeters), but the South American trilobites can be impressively preserved and up to 10 centimeters long.

WANNERIA

Beautifully preserved examples of *Wanneria walcottana* have been known from the Lower Cambrian, Kinzer Formation layers of Pennsylvania for more than 150 years. But it's only been in the last few decades that a so far unidentified species of *Wanneria* has been recognized from the comparably aged Eager Formation of British Columbia.

PARADOXIDES

Perhaps the most noted and obvious example of the Paleozoic phenomenon discussed here exists in the genus *Paradoxides*. The more renowned examples of this trilobite may now be found in Morocco and the Czech Republic, but other prime specimens have been uncovered in Sweden, France, Massachusetts, Spain, Wales, and Newfoundland.

OLENOIDES

The ubiquitous Middle Cambrian trilobite *Olenoides* is best known for its varying species (*O. superbus, O. nevadensis, O. inflatus, O. skabelundi*, and *O. vali*, among others), all of which emerged from the fossil-rich outcrops of Utah. But nearly identical examples appear everywhere from Siberia, to western Canada, to southern China.

OLENELLUS

Few fossilized trilobite genera enjoy a greater worldwide distribution than *Olenellus*. From Scotland,

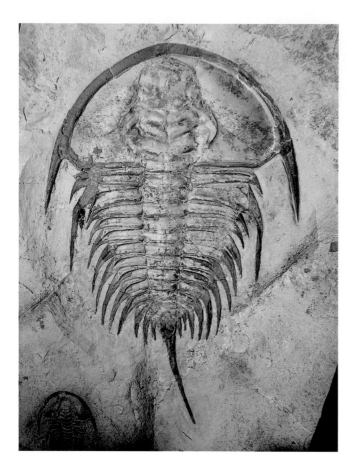

OLENELLUS GILBERTI (MEEK, 1874)
Lower Cambrian; Pioche Shale; Lincoln County, Nevada, United States; 13.5 cm

This is one of the largest O. gilberti ever found.

to Newfoundland, to Pennsylvania, to Nevada, to Alabama, to California, to British Columbia, it is quite apparent that following their 20-million-year swim through the Lower Cambrian seas differing species of this distinctive trilo-type have quite literally been transported (via the power of plate tectonics) to all corners of the globe.

TRIMERUS

From all the evidence presented here, it's apparently not that unusual to discover homogeneous trilobite genera preserved in divergent corners of the planet. But it is rather unusual for each of these examples to share a common species name. So it should be of particular interest to many collectors that specimens of the pervasive homalonotid *Trimerus delphinocephalus* have been reported in New York, England, and Indiana.

REDLICHIA

Among the earliest trilobite genera, some *Redlichia* could grow to impressive sizes, with Australian examples of the recently described *R. rex* reaching 25 centimeters in length. A somewhat smaller but nearly indistinguishable species, *R. chinensis*, is now found some 8,500 kilometers to the north in China's Guizhou Province.

DICRANOPELTIS

Long considered one of the most beautiful trilobites in the world, this rare lichid genus appears in the Silurian-age sedimentary strata of both the Czech Republic and New York State. *Dicranopeltis scabra* was first described by the famed Joachim Barrande in 1853, and *Dicranopeltis nereus* was originally noted by the equally fabled James Hall in 1863.

(LEFT) ECCOPTOCHILE CLAVIGERA (BEYRICH, 1845)

Upper Ordovician; Letna Formation; Beroun, Czech Republic; 6.7 cm

(BOTTOM) ECCOPTOCHILE MARIANA (VERNEUIL AND BARRANDE, 1856)

Upper Ordovician, Caradoc Stage; Andouille Formation; Ponceti Biozone; Mayenne, France; 4.7 cm

***WANNERIA WALCOTTANA* (WANNER, 1901)**

Lower Cambrian; Kinzers Formation; Lancaster County, Pennsylvania, United States; 16.3 cm

WANNERIA SP.

Lower Cambrian; Eager Formation; Cranbrook, British Columbia, Canada; 12.3 cm

(LEFT) *TRIMERUS DELPHINOCEPHALUS* (GREEN, 1832)

Middle Silurian; Waldron Shale Formation; Waldron, Indiana, United States; 13.7 cm

(BOTTOM) *TRIMERUS DELPHINOCEPHALUS* (GREEN, 1832)

Lower Silurian; Rochester Shale Formation; Middleport Quarry; Middleport, New York, United States; 6.5 cm

39

10 LEGENDARY TRILOBITES

T rilobite fossils have existed on our planet for hundreds of millions of years, but the concerted collecting of these primeval relics is a relatively recent occurrence. In the eighteenth century, many Europeans were aware of the "frozen locusts" they occasionally encountered while traversing through their fields. Even earlier, southwestern Native American tribes treated the *Elrathia kingii* carapaces they found with the reverence of religious artifacts, and as far back as the tenth century, noble houses in China proudly presented trilobite fossils as pieces of natural art. Nearly 4,000 years ago, Egyptian royalty flaunted trilobite fossils as symbols of power and prestige. Well-worn and apparently highly prized trilobite specimens have been uncovered among the personal possessions of Ice Age humans discovered in European cave sites dating back more than 15,000 years. However, only in the last 50-plus years has the advent of nimble off-road vehicles, powerful rock-moving machinery, and sensitive GPS trackers made the exploration and excavation of new trilobite-bearing locations accessible to both adventurers and academics. For many fossil enthusiasts, these Paleozoic remnants now rank as the most significant of primordial faunal forms, surpassing even the hallowed dinosaur in their paleontological appeal. Thanks to this intriguing combination of factors—including their easy availability for both study and sale, their often bizarre appearance, and the amazing tales told by their fossilized remains—trilobites, in all their multisegmented glory, represent one of early life's most captivating efforts.

With all of that to consider, here's a salute to 10 legendary trilobites.

PEACHELLA IDDINGSI

For decades, explorers and scientists who traveled to California's formidable Carrara Formation in search of trilobites were confounded by what they uncovered. The Lower Cambrian rocks they extracted often contained small trilobite heads with incredible and disproportionately inflated genal spines. These adventurers were unable to understand the morphological benefit such a feature might possess, and despite their best efforts they couldn't find an intact example of the trilobite species now known as *Peachella iddingsi*. Finally, late in the twentieth century, a small pocket featuring complete specimens of this highly unusual trilobite was discovered.

ASAPHUS KOWALEWSKII

In the last 30-plus years, Ordovician trilobites from Russia have emerged as clear-cut collectors' favorites. With their shiny, three-dimensional, toffee-hued carapaces contrasting attractively against a light-tan matrix, these ancient arthropods feature dozens of intriguing species, many of which have assumed prime positions in both private collections and museum displays. One particular trilobite, the strange but relatively prevalent *Asaphus kowalewskii*—with its eyes sitting atop spindly stalks that frequently reach a length of 5 centimeters or more—has quickly become a "must have" for every arthropod aficionado from St. Petersburg to St. Louis.

ELDREDGEOPS CRASSITUBERCULATA

The Devonian-age *Eldredgeops* specimens found in the famed Sylvania deposits of northern Ohio rank among the best-known of all North American trilobites. Beautifully preserved in a charcoal-gray calcite, these three-dimensional arthropods—which can be collected either tightly enrolled in a defensive position or totally outstretched, and usually range between 4 and 7 centimeters in length—often appear little disturbed by their prolonged passage through evolutionary time. Each of the facets adorning their extraordinary compound eyes is easily observed, and a slight difference in alignment, size, and shape between the lenses distinguishes the two *Eldredgeops* types at this site: *E. milleri* and *E. crassituberculata*. For many collectors, this nominal variance serves as the primary means for both identifying and separating these captivatingly comparable species.

CALYMENE BLUMENBACHII

There was a time in the mid-1800s when group sojourns to the hilly Silurian outcrops of Wren's Nest, located near the town of Dudley, England, represented state-of-the-art academic field pursuits. Pieces of more than 60 trilobite species could be uncovered there, and fossilized fragments frequently littered the ground. Among these scattered remains complete specimens were occasionally found, and most of these became known nationally as the "Dudley Locust"—the trilobite *Calymene blumenbachii*. These discoveries lit a firestorm of scientific interest that quickly spread across the British Isles.

CERAURUS PLEUREXANTHEMUS

During explorations of the fossil-rich Ordovician layers of central New York State in 1871, the legendary Charles Walcott uncovered the initial trilobite specimens definitively exhibiting soft-tissue appendages. News of that unexpected discovery rocked the still nascent paleontological field. Inspired by these uniquely preserved examples of *Ceraurus pleurexanthemus*, Walcott came to the

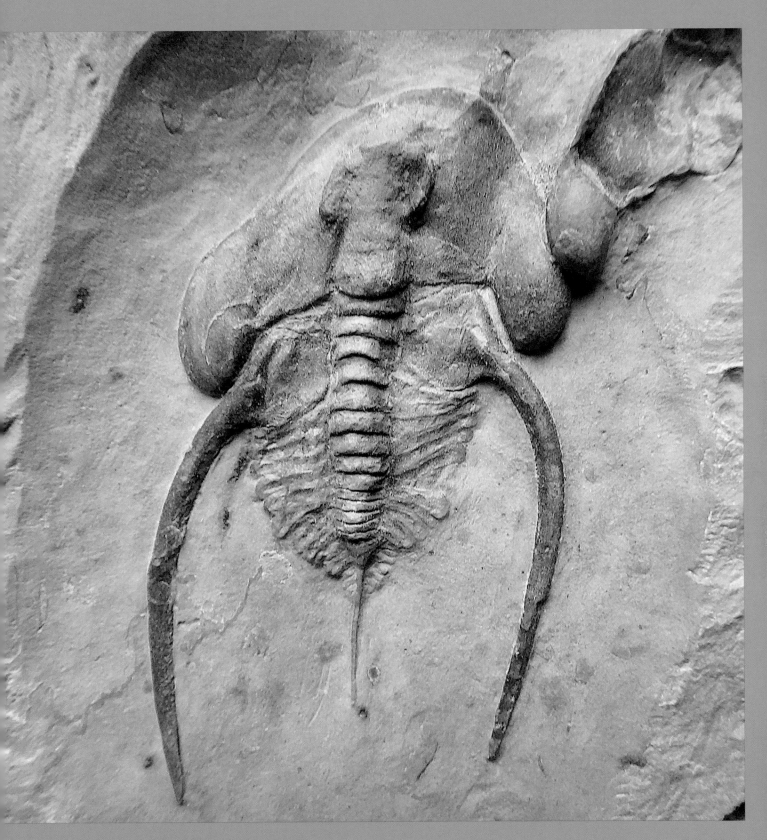

PEACHELLA IDDINGSI (WALCOTT, 1884)

Lower Cambrian; Carrara Formation, Echo Shale Member; Tecopa Hot Springs, California, United States; 3 cm

ASAPHUS KOWALEWSKII LAWROW, 1856
Middle Ordovician, Middle Llanvirnian; Middle Aseri Regional Stage; Duboviki Formation; Volkhov River, St. Petersburg region, Russia; 7.5 cm

fish," which centuries later would be identified as the trilobite *Ogygiocarella debuchii*. In the academic journal *Philosophical Transactions of the Royal Society*, Lhuyd also submitted carefully constructed drawings of his find, which represent the first widely dispersed images of a complete trilobite.

OLENOIDES SERRATUS

There is no more lauded invertebrate fossil repository in the world that the Burgess Shale of British Columbia. In that beautiful outcropping located high in the Canadian Rockies lurk the remains of many notable members of the hallowed Cambrian Explosion. Apart from a plethora of the legendary soft-bodied creatures that inhabited those Middle Cambrian waters (along with the notorious "trilobite eater" *Anomalocaris canadensis*), the most recognized arthropod found at the Burgess locale is the large (up to 10-centimeter-long) trilobite *Olenoides serratus*. Some of these have been revealed with their antennae and walking legs preserved in the site's finely grained mudstone slabs.

realization that he could make a full-time (and profitable) career out of his fossil-centric efforts. As the youthful entrepreneur kept digging, he expanded his scientific base by writing several well-received papers focused on trilobite appendages—the first such works of their kind.

OGYGIOCARELLA DEBUCHII

In England during the late seventeenth century, Reverend Edward Lhuyd made the first direct mention of a trilobite in scientific literature. His description was of something he called a "flat

ISOTELUS MAXIMUS

In just about every major North American museum display, one trilobite stands out from the rest. It may not be the prettiest. It may not be the most intricately designed, nor the most colorful. It may not even be the best preserved. But it's usually the largest arthropod in that public showcase. For more than a century, the impressive (up to 40 centimeter) *Isotelus maximus* specimens that have been pulled from the 447-million-year-old Arnheim Formation layers that sit atop Mount Orab, Ohio, rank among the most recognized and renowned Ordovician trilobite species found in the American Midwest.

CALYMENE BLUMENBACHII DESMAREST, 1817

Middle Silurian, Upper Wenlockian Stage; Much Wenlock Limestone Formation; Wren's Nest Hill, Dudley, West Midlands, England; 6.2 cm

Photo courtesy of the Martin Shugar Collection

BUMASTUS BARRIENSIS

Hailing from one of the preeminent trilobite quarries in the world, the sleek Silurian species *Bumastus barriensis* possesses a long and noble history. Disarticulated pieces of this ubiquitous member of the Corynexochida order were initially noted from the famed Hay Head outcrop in the English Midlands during the early years of the nineteenth century. A few years later, in 1839, Sir Roderick Murchison—one of the most renowned figures in paleontological history—first identified and described this species in scientific literature. Since then a "true" *Bumastus* from Hay Head (to be distinguished from the closely related *Cybantyx* specimens that hail from nearby Wren's Nest) has become a prized possession for those collectors fortunate enough to have acquired a prime example.

DROTOPS ARMATUS

Throughout their prolonged stay on planet Earth, trilobites were forced to protect themselves from a hostile cast of predators—marine menaces that included 50-centimeter-long Cambrian *Anomalocaris*, 60 centimeter Silurian sea scorpions, and giant Ordovician cephalopods. The trilobite line's ability to defend itself apparently reached its apex in the Devonian. At that time in Earth history, Moroccan species such as *Drotops armatus*, which routinely reached lengths of 15 centimeters, had evolved into creatures resembling heavily armored Paleozoic battle tanks—formidable invertebrates covered in dozens of 1- to 2-centimeter spikes. Their fossilized appearance has placed these imposing *Drotops* among the most collectible, legendary, and instantly recognizable of all trilobites.

OGYGIOCARELLA ANGUSTISSIMA (SALTER, 1865)

Lower Ordovician, Llanvirnian Stage; Llanfawr Mudstones Formation, teretiusculus Biozone; Pencerrig Lake (near Builth Wells), Powys, Wales; 7.5 cm; This was the first trilobite genus ever mentioned in literature.

Photo courtesy of the Martin Shugar Collection

(TOP) *BUMASTUS BARRIENSIS* (MURCHISON, 1839)

Middle Silurian, Lower Wenlock Series, Sheinwoodian Stage; Coalbrookdale Formation, Barr Limestone Member; Hay Head, Great Barr, West Midlands, England; 6.3 cm

(BOTTOM) *GLOSSOPLEURA SP.*

Middle Cambrian, Miaolingian Series, Wuliuan Stage; Langston Formation, Spence Shale Member; Antimony Canyon, Wellsville Mountains; Box Elder County, Utah, United States; 5.8 cm

***OLENOIDES SERRATUS* (ROMINGER, 1887)**

Middle Cambrian; Burgess Shale Formation, Campsite Cliff Member; Field, British Columbia, Canada; 6.8 cm

Photo courtesy of the GLC Collection

40

10 MUST-READ TRILOBITE BOOKS

In the last 60-plus years, a variety of notable, and in some cases surprisingly readable, trilobite-related books have made their appearance on the scientific scene. Of course, many of those works (including the legendary *Treatise of Invertebrate Paleontology*, which was released in 1959 but remains an essential part of any serious fossil collector's library) are scholarly tomes written by and for an audience of fellow academics, all of whom seem guaranteed to revel in the morphological minutiae they present. Especially since the beginning of the twenty-first century, a steadily growing stream of books dedicated to highlighting the myriad Paleozoic charms of the world's favorite fossil arthropod have appeared—few of which, somewhat surprisingly, have focused their literary attentions solely on the intellectual whims of the scientific community. These publications have relied primarily on presenting various aspects of the compelling trilobite tale directly to a growing horde of fossil-focused collectors and hobbyists. Some of these captivating efforts have been pocket-sized handbooks written by leading scientists willing to "dumb down" their approach to reach a broader audience. Others are coffee-table-size volumes filled with photos of prime-time trilobite specimens guaranteed to turn any museum curator green with envy. These efforts are almost exclusively the work of amateur enthusiasts whose fascination with these incredible invertebrates has clearly moved past the point of mere interest and now borders precariously on affection, if not outright infatuation.

Here's the *Collector's Guide* list of 10 must-read trilobite books.

TRILOBITES

by Riccardo Levi-Setti

When Ricardo Levi-Setti first released *Trilobites* some three decades ago, it suddenly transformed the planet's understanding and appreciation of these renowned Paleozoic arthropods. With a detailed text that was accessible to the layperson and hundreds of spectacular black-and-white photographs of specimens from throughout the world—drawn from both museum holdings and leading private collections—the book stood at the foundation of a true trilobite renaissance. By introducing both academics and amateur enthusiasts to the then just burgeoning material emerging from Morocco and Russia, as well as featuring first-class specimens from legendary locales throughout North America and Europe, *Trilobites* was a work truly revolutionary in nature.

ORDOVICIAN TRILOBITES OF THE ST. PETERSBURG REGION, RUSSIA

by Arkadiy Evdokimov, V. Klikushin, A. Pilipyuk, and Richard Hightower

Presenting 544 pages filled with nearly 700 color photos of more than 200 different trilobite species, *Ordovician Trilobites of the St. Petersburg Region, Russia* is a feast for both the eyes and the mind. This coffee-table-size volume represents a decade's worth of cumulative effort by the book's four authors. From the most common asaphid to the rarest lichid, they're all here in their surprisingly spinose glory—complete, beautifully prepared specimens that stand in sharp contrast to the often fragmentary material frequently featured in scientific literature and scholastic works. Indeed, this book will quickly prove useful to both the trilobite-obsessed collector and the most high-minded

academic. Unquestionably, despite its high cost, *Ordovician Trilobites of the St. Petersburg Region, Russia* stands among the true must-have volumes for anyone even remotely interested in trilobites.

TREATISE OF INVERTEBRATE PALEONTOLOGY, PART O: ARTHROPODA, VOL. 1

The legendary, lauded, and now often difficult to obtain *Treatise of Invertebrate Paleontology* remains an essential ingredient of any comprehensive trilobite library. Written by many of the world's leading mid-twentieth-century scientific minds (and edited by Raymond C. Moore of the University of Kansas), upon its release in 1959, this volume—which exclusively features trilobites along with what were then perceived to be soft-tissue "trilobitomorphs"—gave many future academics and collectors their first real taste of arthropod-associated knowledge. Much of the information is now clearly out of date, and many of the specimens—featured in both grainy black-and-white photographs and line drawings reflecting various degrees of complexity—have been renamed or reclassified, but the treatise is still an important reference tool for anyone fascinated by the trilobite realm.

TRILOBITES OF THE BRITISH ISLES

by Robert Kennedy and Sinclair Stammers

The broad swath of trilobite material that graces this diminutive European kingdom is celebrated in the softcover tome succinctly titled *Trilobites of the British Isles*. Written by the noted trilobite collector/researcher Robert Kennedy, with the able assistance of the photographer Sinclair Stammers, this highly informative, richly illustrated, 384-page volume goes to laudatory lengths in its attempts to present

KOOTENIA SPENCEI RESSER, 1939
Middle Cambrian, Miaolingian Series, Wuliuan Stage; Langston Formation, Spence Shale Member; Antimony Canyon, Wellsville Mountains; Box Elder County, Utah, United States; 6.4 cm

as all-encompassing a work as paleontologically possible. To achieve his goal of updating more than 300 years of British trilobite tradition and research, Kennedy has not only focused on presenting the stellar arthropod attractions housed in his own world-class collection but has also delved deep into the resources provided by his fellow trilobite enthusiasts throughout the United Kingdom. He also presents some incredible specimens long hidden behind the imposing walls of leading British museums.

TRILOBITES OF NEW YORK: AN ILLUSTRATED GUIDE

by Thomas Whiteley, Gerald Kloc, and Carlton Brett

Even for the most fossil-obsessed among us, when thoughts of New York come to mind, it's safe to say

that rarely are those mental images of trilobites. If any book has a fighting chance of changing that Big Apple perspective, however, it's *Trilobites of New York: An Illustrated Guide.* Throughout this voluminous, academically inclined work, the three authors present the diverse and beautiful trilobite fauna of the Empire State with an unmistakable passion and a welcomed degree of perception. New York's fascinating paleontological history—featuring the likes of James Hall and Charles Walcott—is showcased along with detailed descriptions of the area's key fossil localities. There are also revealing black-and-white photos of the unquestioned stars of the show, the state's dizzying array of trilobites. More than 200 species are highlighted, amazing creatures that crawled through the region's rich Paleozoic seas from the Cambrian through the Devonian—an impressive stretch of 170 million years.

TRILOBITES OF THE WORLD: AN ATLAS OF 1000 PHOTOGRAPHS

by Pete Lawrance and Sinclair Stammers

With *Trilobites of the World: An Atlas of 1000 Photographs*, author Pete Lawrance and photographer Sinclair Stammers have carefully assembled a pleasing and easily digestible visual display of information and images. As the title indicates, an impressive assortment of 1,000 color photos depicting more than 700 species are featured. This is a compact, well-designed softcover volume that deftly delves through the nearly 300-million-year undersea reign of these amazing arthropods. Lawrance is a leading collector and Stammers is a renowned photographer. The pair make a dynamic duo when it comes to gathering together and presenting specimens from the four corners of the globe. Highlights include beautiful and unusual

**NIOBELLA HOMFRAYI
(SALTER, 1866)**

Lower Ordovician; Tremadoc Series;
Penmorfa Church (near Portmadoc),
Gwynedd, Wales, United Kingdom;
17.2 cm

trilobites from Great Britain (no surprise considering that both authors have British roots) as well as recently uncovered material from paleontological hotbeds such as Morocco and Russia. Also featured are eye-catching examples of unusual species from Bolivia, Canada, Germany, and China.

ORDOVICIAN TRILOBITES OF SOUTHERN ONTARIO, CANADA AND THE SURROUNDING REGION

by Phillip A. Isotalo

In his illustrated guide *Ordovician Trilobites of Southern Ontario, Canada and the Surrounding Region*, author Phillip A. Isotalo presents the best known—as well as many of the often overlooked—species from that fossil-filled part of the planet. He does so in an eye-catching and informative manner designed to appeal to the hearts and minds of trilobite enthusiasts everywhere. Featuring a captivating display of detailed color photos, revealing charts, and comprehensive descriptions of such notable trilobites as *Gabriceraurus dentatus, Isotelus gigas*, and *Hemiarges paulianus*, this concise, 224-page, softcover volume effectively exhibits the incredible diversity of species that appear in the region's various Ordovician-age formations, outcrops, and quarries.

BACK TO THE PAST MUSEUM GUIDE TO TRILOBITES

by Carlo Kier and Enrico Bonino

Since its release in 2010, *Back to the Past Museum Guide to Trilobites* has been a warmly embraced addition to the available resources focused on these long-departed marine inhabitants. Beautifully designed and lavishly illustrated, this nearly 500-page self-published volume—a joint project of the noted collector Carlo Kier and the scientist Enrico Bonino—is a joy to peruse even though its sheer size and weight makes it challenging merely to hold and navigate. Divided into sections that detail subjects such as Trilobite Morphology and Classification before getting "down to business" with more than 400-full-color-photo plates presenting eye-popping specimens ranging from Cambrian to Carboniferous, this edition seems determined to make every museum curator take notice and every trilobite collector drool with envy. Delving deeply into famed trilobite deposits in China, Italy, Bolivia, Morocco, and the United Sates (among others), this is one book truly worldwide in its scope.

TRILOBITE! EYEWITNESS TO EVOLUTION

by Richard Fortey

During the last four decades, Richard Fortey has emerged as one of the contemporary scene's most compelling and significant academic voices, especially when the subject happens to be trilobites. Long a world-renowned paleontologist at London's famed Natural History Museum, in recent years Fortey has increased his public profile through regular television appearances and by writing a series of well-received books, including *Trilobite! Eyewitness to Evolution*. Through an engrossing blend of easily relatable, scientifically based stories, personal accounts, and "behind the scenes" museum musings, Fortey crafts an entertaining and frequently captivating tale of the trilobites' incredible ride through the Deep Time seas. Although virtually devoid of the kind of eye-popping photos that make the average trilobite fan's heart beat just a little bit faster, this is a "must-have" for any comprehensive natural history library.

TRAVELS WITH TRILOBITES: ADVENTURES IN THE PALEOZOIC

by Andy Secher

I admit it, at times I can be the epitome of self-serving. But at least I didn't put my *favorite* trilobite-related effort at the top of this invertebrate-loving top 10 list! As the often imitated, never duplicated precursor to the *Trilobite Collector's Guide* currently in your possession, *Travels with Trilobites* attempts to literally and figuratively go where no trilobite book has gone before. In stark contrast to the plethora of textbooks, dissertations, and treatises that have previously addressed the subject of these amazing arthropods, this 400-page, hardcover volume (with nearly 300 incredible color photos) is a rollicking read that manages to blend fact, fun, and fossiliferous content into a package of pure Paleozoic infotainment. As the title implies, *TwT* takes readers on a worldwide ride in search of the best trilobites and the most fascinating locations on the planet, and I feel confident in saying that it succeeds on both counts. For obvious reasons, I *highly* recommend it!

(LEFT) *DIACALYMENE DRUMMUCKENSIS* **REED, 1906**

Upper Ordovician, Ashgillian Stage, Rawtheyan Substage; South Threave Formation, Lower Drummuck Mudstones Member; Girvan, Ayrshire, Scotland; 6.5 cm

(BOTTOM) *REDLICHIA MANSUYI* **RESSER AND ENDO, 1937**

Lower Cambrian, Series 2, Duyunian; Guanshan Fauna; Wulongqing Formation; Caijiachong Valley, Kunming City, Yunnan Province, China; 7.4 cm

LIEXIASPIS ZHENGJIAENSIS YIREN, 1982

Lower Ordovician, Upper Tremadocian; Fengxiang Formation; Liexi, Hunan, China; 8.2 cm

SELENOPELTIS AFF. BIONODOSA DEAN, 1966

Lower Ordovician, Arenigian; Upper Fezouata Formation; Zagora, Morocco; 5.7 cm

41

10 STRIKINGLY STRANGE TRILOBITES

Trilobites were among nature's most adaptable creatures, surviving in an ever-changing marine ecosystem for more than a quarter-of-a-billion years. Despite constant threats, more than 25,000 species of these ancient arthropods are now known to have inhabited Earth's primeval seas. Amid this mind-boggling degree of biological diversity are some trilobite species that are so unusual, so strange, and so brazenly bizarre that they merit special morphological mention. These creatures are so alien in appearance that they could easily be featured attractions in any upcoming Hollywood sci-fi spectacular—but few observers would believe that such life-forms could possibly have existed. But they did, and they were among the first residents of Earth's Deep Time seas. The heterogeneity of trilobite body designs is almost impossible to fathom, even for those who have long collected and studied the fossilized remains of these primeval ocean inhabitants. Amid those assorted anatomical alignments are some surrealistic features that even the likes of Picasso would have had difficulty envisioning—and those features are often attached to trilobites that collectors around the globe find themselves drawn to like the metaphorical bees to honey.

Here's a brief look at 10 of the strangest trilobites ever to appear in Earth's waters.

SPATHACALYMENE NASUTA

Trilobite collectors can be a bewildering bunch. No matter how many specimens they own or how many unique species their Paleozoic assemblage represents, another bug is always ready to top their insatiable "want" list. For many hobbyists around the globe, the instantly recognizable *Spathacalymene nasuta* from the Silurian-age Osgood Formation outcrops of Indiana is near the pinnacle of any such listing. With its pronounced cephalic rostrum, lovely caramel-colored shell, and pronounced three-dimensional preservation, it's not surprising that so many collectors crave a top-quality "spath" to add to their fossiliferous holdings. Considering that no more than 50 complete specimens have ever been uncovered, the hunt for that prime example may prove to be longer and more difficult than initially expected.

ASAPHUS KOWALEWSKII

When 450-million-year-old Russian trilobite fossils first began to invade the world market in the early 1990s, one species stood above the rest—both literally and figuratively. With its optics perched atop 5-centimeter-long stalks, *Asaphus*

kowalewskii truly resembled a creature from a distant galaxy. As it turned out, that eye-popping anatomical feature must have been quite advantageous because hundreds of complete fossilized examples of this bizarre trilo-type have emerged from the quarries adjacent to St. Petersburg. Their abundance has made these distinctive Ordovician trilobites an essential component of any *au courant* museum display or major private collection.

BRISTOLIA INSOLENS

Until late in the twentieth century, the primitive species *Bristolia insolens* was recognized almost exclusively from the disarticulated exuviae of its distinctively shaped, semirectangular cephalon. These unusual trilobites—known primarily from fossilized remains found in the 520-million-year-old Latham Shale of California and the Pioche Shale of Nevada—could grow to impressive dimensions, with some complete examples topping 6 centimeters in length. With a sweeping pair of genal spines bordering the sides of its head, this Lower Cambrian species vividly displays the trilobite class's incredible diversity of form very early in their 270-million-year swim through Deep Time.

(OPPOSITE PAGE) **ODONTOCEPHALUS AGERIA (HALL, 1861)**

Middle Devonian; Onondaga Limestone; Needmore Shale Member; Perry County, Pennsylvania, United States; larger specimen 11.2 cm

WALLISEROPS TRIFURCATUS MORZADEC, 2001

Lower Devonian, Upper Emsian; Timrhanrhart Formation; Foum
Zguid, Southern Morocco; 7.2 cm

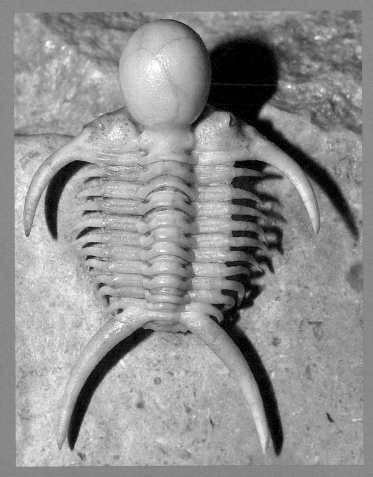

SPHAEROCORYPHE DENTATA ANGELIN, 1854

Upper Ordovician, Sandbian; Alekseevka Quarry; St. Petersburg,
Russia; 2.3 cm

Photo courtesy of the Sam Stubbs Collection

(OPPOSITE PAGE) **PROBOLICHAS KRISTIAE
CARLUCCI, WESTROP, AND AMATI, 2010**

Upper Ordovician; Bromide Formation, Pooleville Member;
Arbuckle Mountains, Oklahoma, United States; 9.3 cm

(*TOP, LEFT*) CALYMENE SP. AND (*BOTTOM, RIGHT*) SPATHACALYMENE NASUTA (ULRICH, 1879)

Middle Silurian, Sheinwoodian; Osgood Formation; Ripley County, Indiana, United States; *Spath* 6.3 cm, *Caly* 5.8 cm

While Spaths are rare, this site's Calymene examples are even rarer.

42

10 TOP TRILOBITE WEBSITES

From the dawn of human history until an indeterminate moment in the "primal" past—say, the turn of the most recent millennium—the only sure way to get your hands on a trilobite was to visit a rock and mineral show, trade for it with a like-minded acquaintance, or venture into the field and collect it yourself. Since those distant yesterdays, access to quality fossil material has changed radically. No longer do you have to plan a journey by foot, car, train, or plane to see and acquire the latest and greatest in trilobite treasures. Indeed, all you now need to do is sit in a cushy chair in the comfort of your own home and browse through your phone, iPad, or laptop. In the ever-expanding sphere of internet experiences, an entire panoply of incredible fossil collecting opportunities is instantly available to both novice and advanced consumers of natural history goods. Dozens of fossil-oriented websites now provide even the most advanced arthropod aficionados with the chance to purchase eye-catching trilobite specimens found throughout the world—at price points ranging from pocket change to profoundly big ticket. And never forget the online monster in the room: the ubiquitous eBay. Literally thousands of trilobites—from the most common species to the most exotic—are available to anyone, anywhere, on an unrelenting 24/7 basis. Community sites such as Facebook and Instagram also provide potential access to top-grade trilobite material as they bring interested collectors into direct contact with tens of thousands who share their Paleozoic passion. Whether you are assembling your own museum-worthy trilobite collection or merely looking for a single top-grade example to show

***PLATYLICHAS GRAYI* REED, 1906**
Upper Ordovician, Ashgillian Stage, Rawtheyan Substage; Upper Drummuck Group, South Threave Formation, Farden Member; Starfish Bed, Lady Burn, Girvan, Ayrshire, Scotland; 2.7 cm

friends and family, there are many ways to locate trilobites using the internet.

Here's a list of 10 top trilobite websites, including a few of the most influential and most frequently visited noncommercial outlets.

EBAY

The biggest, boldest purveyor of trilobite goods in the natural history universe is eBay. It can be either a blessing or a curse, depending on your paleontological point of view. Thousands of trilobite specimens appear on the site at any given moment, most relatively common or poorly preserved examples from Morocco, Russia, or North America. But occasionally spectacular and rare pieces do come up for auction, a fact that keeps the trilo-centric among us continually returning to

this unpredictable, often derided, all things to all people website.

FOSSILERA

The new "king" of the online natural history scene since its inception in 2017, FossilEra has solidified its position among the most viewed and successful fossil-oriented outlets on the internet. Some of their featured trilobites were self-collected—such as exciting Ordovician material from Utah's Fillmore Formation—and this site mixes an overwhelming barrage of decorator-style minerals, carved goods, and dinosaur teeth with an intriguing blend of trilobites that run the price gamut from reasonable to outrageous.

FOSSIL MALL

There have been moments in the last two decades when this cooperative site—which presents a revolving list of paleontologically inclined participants—has featured some of the major "players" in the trilobite world; their corresponding listings have been quickly perused and purchased by many of the planet's top collectors. At other times, however, the site's erstwhile stars have gone on temporary or permanent hiatus, leaving Fossil Mall sorely lacking in must-see material. The site is worth checking out, but your level of success and enjoyment may depend on which day of the week (or year) you choose to visit their domain.

PALEOART

When the topic is Russian trilobites, the first stop on any serious collector's quest has long been this elegantly designed website. For more than a decade, Paleoart has been presenting the crème de la crème of Ordovician material emerging from quarries that circle the St. Petersburg vicinity. Their prices can be jaw-dropping, although occasionally negotiable, and the constant flow of new material (which sometimes also features Lower Cambrian species from Siberia) is more than enough to have trilobite aficionados from around the globe make this a regular stop in their internet pursuits.

EXTINCTIONS

In the 1990s, during the nascent days of the internet, few natural-history-oriented sites made more of an impact on the hearts, minds, and bank accounts of collectors than Extinctions. Their weekly and monthly trilobite auctions were circle the calendar, can't miss events for fossil lovers around the globe. In recent years, a bit of the bloom has come off the Extinctions rose, and their Paleozoic sales now revolve more around well-prepared but no longer super rare species from Morocco, Russia, and North America.

FACEBOOK

This controversial online community seems to swing in and out of favor with the main purveyors of contemporary coolness. One fossil-focused Facebook page that provides a continually updated "all things trilobite" forum recently bragged about attaining an active community of more than 10,000 members—perhaps not quite up to Cristiano Ronaldo standards (that soccer dude has more than 150 million followers) but still rather impressive considering its less than mainstream subject matter. Like it or loathe it, Facebook has come to represent a state-of-the-art way to reach out to those who share your trilo-centric fascination.

THE NATURAL CANVAS

The first thing most arthropod-obsessed internet searchers may note when visiting the well-designed

***ACASTELLA HEBERTI ELSANA* RICHTER AND RICHTER, 1954**
Topmost Silurian, Basal Devonian transition; Pridolian/ Lochkovian Stage sequence; Rukshin Tier (Regional Series); Skala Formation, Dniester River section; Dnistrove, Ternopil Oblast, Ukraine; 3.2 cm

and easy to navigate Natural Canvas site is the surprising number of high-grade (and highly priced) examples of rare Burgess Shale fossils being offered for sale. Acquiring such Middle Cambrian material has long been exceedingly difficult—and in some cases highly questionable because the Canadian government continues to maintain a vicelike grip on virtually all Burgess-derived paleo-matter—but this Texas-based firm seems to know how to skirt around such potentially ticklish issues. The Natural Canvas has been offering top-end Burgess stock, along with a smattering of rare trilobites from around the world, for the better part of two decades.

EVOLUTION2ART

Most of the leading natural-history-oriented websites are located in the United States, but some of the best are now firmly planted on the other side of the Big Pond. In all honesty, one of the inherent beauties of the internet is that it really doesn't matter where your business is situated—except, of course, when it comes to the shipping costs incurred for any purchased material. The preponderance of trilobite specimens offered by Evolution2Art tend to be well-prepared examples of relatively common species, but this stylish site sometimes features amazing European and North American trilobites, along with the by now customary high-end pieces from Morocco and Russia.

AMNH TRILOBITE WEBSITE

In the spirit of full disclosure, I freely admit that I write and provide virtually all the photographed specimens that appear on the popular trilobite website of the American Museum of Natural History. But that thinly guarded revelation doesn't necessarily mean that I am incapable of offering an objective opinion. The AMNH site is among the best noncommercial, fossil-oriented outlets on the internet. It is designed to be a major resource for collectors, students, and researchers around the globe. Photos of more than 1,000 trilobites are presented—some representing unique species from distant and not so distant corners of the globe—along with detailed descriptions of many of the world's top trilobite-bearing locations, ranging from Emu Bay, Australia, to Wren's Nest, England, to Middleport, New York.

TRILOBITES.INFO

This site is the singular vision of amateur enthusiast Sam 'Ohu Gon, senior scientist and cultural advisor for the Nature Conservancy of Hawai'i. In the last two decades, the noncommercial Trilobites.info site has emerged as perhaps the most visited and influential internet address for those interested in Paleozoic arthropods. Filled with a heady blend of cutting-edge scientific research, revealing photos, and thought-provoking trilo-centric information (including detailed sections on Paleobiology and Morphology), this site effectively walks the delicate line between the purely academic and the empirically entertaining (although Gon's approach clearly favors the former). Posting of the site's Trilobite of the Month photo has become an eagerly anticipated event in trilobite circles, and collectors around the globe compete to have their favorite specimen chosen for that singular honor.

(OPPOSITE PAGE) *HARPIDES SP.*
Lower Ordovician, Upper Tremadocian; Fezouata Formation; Zagora, Morocco; 6.3 cm

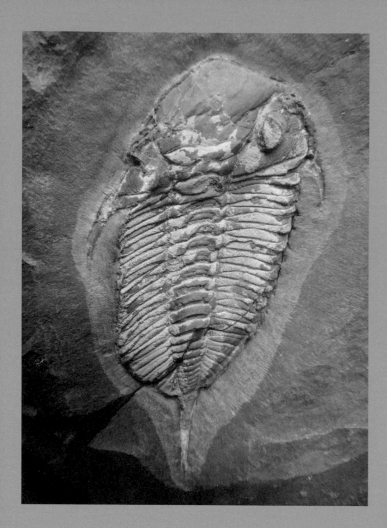

(LEFT) **DALMANITINA RABANOAE** CARDOSO PEREIRA, 2017

Upper Ordovician, Sandbian; Ribeira do Cavalinho Formation; Mação, Portugal; 4.5 cm

(BOTTOM) **BICERATOPS NEVADENSIS** PACK AND GAYLE, 1971

Middle Cambrian, Miaolingian Series, Delamaran Regional Stage; Bright Angel Shale; Frenchman Mountain, Nevada, United States; 5 cm

43

10 UNBELIEVABLE UTAH TRILOBITES

ore than 500 scientifically recognized trilobite species have been described from Utah's fossil-rich sedimentary strata. Many of these Paleozoic remnants are indigenous only to this region's densely packed limestone repositories. The state's principle geological exposures include the Middle Cambrian Wheeler, Marjum, and Weeks formations, as well as the Lower Ordovician Fillmore and Wahwah outcrops. Three orders of trilobites, featuring scores of genera and hundreds of distinct species (such as *Athabaskia gladei*, *Olenoides pugio*, and *Bathyuriscus fimbriatus*), have been identified from the state's Cambrian layers alone, along with a variety of worms, lobopods, and giant soft-bodied arthropods. Trilobite collecting in Utah enjoys a long and fascinating history. Indeed, Native American petroglyphs that seemingly depict trilobites have been found adorning cliff walls in the state's southern sector, and these somewhat fanciful images could be hundreds, if not thousands, of years old. Among the prominent paleontologists who searched for these primordial relics throughout Utah's rugged landscape was the seemingly omnipresent Charles Walcott. Late in the nineteenth century he frequently visited this western outpost in quest of new trilobite-bearing exposures. When all things are considered, it should be of little surprise that this majestic state enjoys the distinction of being the most bountiful trilobite storehouse on the North American continent.

Here's a brief look at 10 notable Utah trilobites drawn from a variety of locations and time periods.

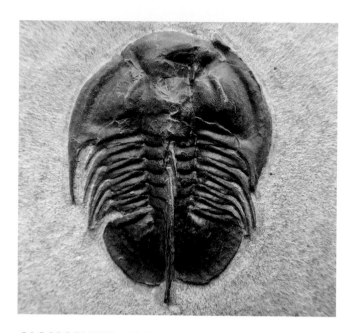

GLOSSOPLEURA YATESI ROBISON AND BABCOCK, 2011

Middle Cambrian, Miaolingian Series, Wuliuan Stage; Langston Formation, Spence Shale Member; Box Elder and Cache counties, Utah, United States; 4 cm

GLOSSOPLEURA YATESI

In a mountain-strewn state such as Utah, merely reaching many of the region's most significant trilobite-bearing deposits can prove to be a long and challenging affair, but the results are often well-worth the effort. Such is certainly the case for those pursuing the Middle Cambrian species *Glossopleura yatesi.* Found in the prolific yet remote environs of the high-altitude Spence Shale—where some of the richest Paleozoic deposits sit at breathtaking elevations of over 2,800 meters—this species has long been known but was only officially recognized in a 2015 academic paper that addressed many of the state's previously unidentified trilobites. With its pronounced axial spines, it is easy to distinguish *Glossopleura yatesi* from closely aligned species such as *G. bion* and the ever-popular collector's favorite *G. gigantea.*

ATHABASKIA GLADEI

The diversity of trilobite external design is astonishing. During their 270-million-year journey through evolutionary time, these incredible invertebrates generated almost 5,000 different genera and more than 25,000 recognized species. Some were covered in spines, others were as sleekly hydrodynamic as a twenty-first-century submarine. Still others, such as *Athabaskia gladei,* a Middle Cambrian species from Utah, possessed a flat, elliptical configuration somewhat reminiscent of a Paleozoic pancake, although it must be acknowledged that some of this specimen's fossiliferous appearance is directly due to millions of years of compression in Earth's ever-shifting sedimentary strata. Such specialized morphology indicates a life lived along the ancient ocean bottom where *Athabaskia* could safely scrounge for sustenance amid the scattered seabed debris.

UTASPIS MARJUMENSIS

Often found amid the same 510-million-year-old Marjum Formation layers that produced both *Modocia typicalis* and the rare *Olenoides pugio,* the attractive *Utaspis marjumensis* represents a highly distinctive Middle Cambrian faunal element. Perhaps the most notable physical characteristics displayed by these graceful trilobites are their gently flaring genal spines and the wide disparity of sizes they could attain. A significant number of recovered examples are 4 to 6 centimeters in length, but some complete *Utaspis* have been found that are 12 centimeters long. The disarticulated fragments—with more than 90 percent of the specimens being molts lacking their free cheeks—indicate that some members of the species may have grown even larger.

ATHABASKIA GLADEI ROBISON AND BABCOCK, 2011
Middle Cambrian, Miaolingian Series, Wuliuan Stage; Langston Formation, Spence Shale Member; Wellsville
Mountains; Box Elder and Cache counties, Utah, United States; 6.7 cm

MODOCIA TYPICALIS

If the ubiquitous *Elrathia kingii* does not symbolize the trilobites of Utah to fossil enthusiasts from Salt Lake to Lake Geneva, then the far rarer but ever-popular *Modocia typicalis* may be the species most qualified to assume that distinction. Perhaps the most intriguing characteristic of this midsized (4- to 8-centimeter) species is the astonishing variety of calcified colors in which it appears. Beautifully preserved examples of *Modocia* have been revealed in Middle Cambrian outcrops throughout the state. They are fossilized in vivid tones of black, red, brown, and tan—each of which then contrasts against a host shale that ranges from gray, to beige, to pink. In combination, the resulting specimens rank among the most appealing (and collectible) in the entire trilobite realm.

TRIGONOCERCELLA ACUTA

Trilobite fragments, which can indicate specimens of particularly notable size or rarity, are common finds throughout Utah's Middle Ordovician Fillmore Formation. Ironically, at times these disarticulated trilo-bits provide better evidence of the depth and diversity of a region's trilobite population then the complete arthropod examples so coveted by collectors around the globe. Impressive pieces of the strangely shaped *Trigonocercella acuta* have been found throughout the Fillmore since collecting began at the site more than a century ago. But only in recent years have complete examples of this distinctive species been recovered, some up to 9 centimeters in length. A little careful prep work occasionally reveals trilobites of virtually unmatched elegance and beauty.

LEMUREOPS LEMUREI

Utah is world renowned for its incredible multiplicity of trilobite-bearing Cambrian deposits, but the state also features a variety of significant Ordovician-age outcrops. These include the closely aligned Fillmore and Wahwah formations, both of which appear as thinly banded layers throughout the region. Dozens of beautifully preserved trilobites have been pulled from the surrounding sites since commercial collecting in these repositories began late in the twentieth century. One of the more noteworthy species to emerge from those digs is *Lemureops lemurei*, a distinctive, somewhat triangularly shaped trilobite that until recently had been assigned to the genus *Pseudocybele*.

MENISCOPSIA BEEBEI

Of all the celebrated Weeks Formation trilobites (and for those wondering, more than two dozen species have been described from that Middle

Cambrian outpost, with more examples apparently on the impending docket for academic analysis and acknowledgment), few are more instantly recognizable than *Meniscopsia beebei*. With its moonlike morphological form, this rare and often beautifully preserved species—which underwent a twenty-first-century change in nomenclature—frequently represents the highlight of any western-centric arthropod assemblage. In some fossilized examples, the gut tract has been preserved, indicating to some academics that this species may have had a predatory nature.

BATHYURISCUS FIMBRIATUS

One of the true trilobite treasures drawn from the Middle Cambrian rocks of Utah is the classic ovate species *Bathyuriscus fimbriatus*, examples of which range between 4 and 9 centimeters in length. Although rare in complete form (the vast majority of recovered specimens are molts missing their free cheeks), even most midsized fossil collections will feature a nicely detailed, representative example of this easily recognized trilobite. Collectors fortunate enough to possess a complete, well-preserved, articulated specimen may note the series of fine lines that characteristically radiate across this species' free cheeks.

ZACANTHOIDES GRABAUI

From the momentous moment they first emerged in the Cambrian seas some 521 million years ago, evidence of trilobite heterogeneity has been pervasive. One of the most intriguing early examples of the trilobite line was the instantly recognizable *Zacanthoides grabaui*, a Middle Cambrian species found in Utah's Spence Shale featuring a triangular cephalon that came to a rather pronounced point. There is still a hot (well, perhaps lukewarm) debate among certain arthropod-obsessed enthusiasts when it comes to deciphering exactly what role such an unusual cephalic shape played in this trilobite's daily battle for survival in those challenging Paleozoic seas. Current academic thought is that this graceful creature may have lived a pelagic lifestyle riding along on gentle ocean currents.

ELRATHIA MARJUMI

When most Paleozoic hobbyists consider the genus *Elrathia*, the ever-popular *E. kingii* quickly and rightfully springs to mind. After all, that is the most commonly recognized trilobite, not only in Utah but in the entire fossil domain! But other species of *Elrathia* are lurking in the rugged outcrops of this mountainous western state, perhaps most notably the far rarer *Elrathia marjumi*. This creature is noted both for its rather unexpected location of discovery in the Marjum Formation (*E. kingii* is exclusively found in the closely aligned Wheeler Shale) and for its slight morphological variances, which include small pleural spines emanating from two of its thoracic segments.

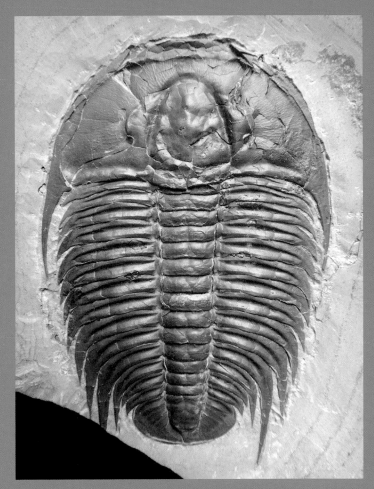

(TOP, LEFT) *UTASPIS MARJUMENSIS* **(RESSER AND ROBISON, 1964)**

Middle Cambrian; Marjum Formation, House Range; Millard County, Utah, United States; 6.1 cm

(TOP, RIGHT) *MODOCIA TYPICALIS* **(RESSER, 1938)**

Middle Cambrian, Series 3, Drumian; Lower Marjum Formation; House Range, Utah, United States; 7 cm

(RIGHT) *LEMUREOPS LEMUREI* **(HINTZE, 1953)**

Lower Ordovician; Fillmore Formation, Pogonip Group; Millard County, Utah, United States; 3.6 cm

The head is slightly tucked on this exceptionally large example.

(OPPOSITE PAGE) *OLENOIDES PUGIO* **WALCOTT, 1908 AND 2** *UTASPIS MARJUMENSIS* **(RESSER AND ROBINSON, 1964)**

Middle Cambrian; Marjum Formation, House Range; Millard County, Utah, United States; *Olenoides* 6.4 cm

(TOP) **MENISCOPSIA BEEBEI ROBISON AND BABCOCK, 2011**

Upper Middle Cambrian; House Range; Weeks Formation; Millard County, Utah, United States; 2.8 cm

(RIGHT) **OLENOIDES SP.**

Middle Cambrian; Upper Wheeler Formation; Drum Mountains, Utah, United States; 12.2 cm

44

10 TRILOBITE ORDERS

During their 270-million-year journey through evolutionary time, trilobites generated more than 180 academically accepted families—an impressive number featuring nearly 5,000 genera and 25,000 distinct species. Their almost unimaginable longevity and multiplicity has continually presented paleontologists with an intimidating yet fundamental challenge: how to best categorize and distinguish one group of trilobites from another. The issue boils down to finding the most expeditious and "orderly" manner of classifying these primordial creatures, one that allows us to gain at least a fundamental understanding of which families, genera, and species should be grouped together and which subsequently produced a logical line of descendants. Some trilobite orders, such as the Lichida, arose in the Upper Cambrian, existed throughout the Ordovician and Silurian, and ended their slither through the Paleozoic in the Devonian—a span of 180 million years. Then consider the Proetida; this order included thousands of species, produced members that first emerged in the Middle Cambrian and lasted all the way to the demise of the entire trilobite class at the end of the Permian—an impressive stretch of 250 million years. In dramatic contrast is the highly important early order Redlichiida: it arose, peaked, and vanished all within roughly a 20-million-year span of the Cambrian. Through the early years of the twenty-first century, it was generally accepted that there were nine distinct trilobite orders—the groupings into which every trilobite can be placed in some sort of basic evolutionary pattern. These orders were Proetida, Asaphida, Phacopida, Lichida, Ptychopariida, Harpetida, Corynexochida, Redlichiida, and Odontopleurida. In 2020 a tenth order,

CHASMOPS MARGINATUS SCHMIDT, 1881
Middle Ordovician; Asery Level; Volkhov River region, St. Petersburg, Russia; 5.3 cm

ORDER: Phacopida

Trinucleida, was proposed that would include some species previously assigned to the Asaphida line.

In keeping with this book's well-defined theme, here is a list with 10 representative species drawn from each of the 10 trilobite orders.

PROETIDA (MIDDLE CAMBRIAN–UPPER PERMIAN)

Basidechenella rowi, Ameura major, Ditomopyge producta, Pudoproetus fernglenensis, Phaetonellus planicauda, Basidechenella lucasensis,

Raymondites spiniger, Xiphogonium trautensteinensis crassus, Cordania wessmani, Bollandia globiceps

ASAPHIDA (MIDDLE CAMBRIAN–UPPER ORDOVICIAN)

Isotelus latus, Asaphus kowalewskii, Dikelocephalus minnesotensis, Trigonecerca piochensis, Megistaspis triangularis, Homotelus bromidensis, Ogygiocarella augustana, Asaphopsoides brevica, Hypodicranotus striatulus, Orygmaspis spinula

PHACOPIDA (LOWER ORDOVICIAN–UPPER DEVONIAN)

Placoparina sedgwicki, Gabriceraurus dentatus, Huntoniatonia lingulifer, Greenops whiteleyi, Glyptambon verrucosus, Eldredgeops rana, Trimerus delphinocephalus, Balizoma variolaris, Pliomera fischeri, Cheirurus ingricus

LICHIDA (UPPER CAMBRIAN–UPPER DEVONIAN)

Allolichas halli, Metopolichas verrucosus, Acanthopyge haueri, Autoloxolichas laxatus, Damesella paronai, Selenopeltis buchii, Taihungshania miqueli, Dicranopeltis scabra, Amphilichas ottawaensis, Belenopyge balliviani

PTYCHOPARIIDA (LOWER CAMBRIAN–UPPER ORDOVICIAN)

Conocoryphe sulzeri, Modocia typicalis, Tricrepicephalus texanus, Ellipsocephalus hoffi, Elrathia kingii, Estaingia bilobata, Amecephalus laticaudum, Weeksina unaspina, Triarthrus eatoni, Asaphiscus wheeleri

HARPETIDA
(MIDDLE CAMBRIAN—UPPER DEVONIAN)

Boheomoharpes ungula, Harpides grimmi, Aristoharpes perradiatus, Scotoharpes spasski, Dolichoharpes reticulatus, Hibbertia ottawaensis, Eoharpes benignensis, Australoharpes cordilleranus, Paraharpes hornei, Lioharpes crassimargo

CORYNEXOCHIDA
(LOWER CAMBRIAN—MIDDLE DEVONIAN)

Olenoides superbus, Marjumia typa, Paralejurus hamlagdadicus, Bumastus ioxus, Cybantyx anaglyptos, Thysanopeltis speciosa, Orytocephalus palmeri, Eobronteus lunatus, Zacanthoides typicalis, Hemirhodon amplipyge

REDLICHIIDA
(LOWER CAMBRIAN—MIDDLE CAMBRIAN)

Elliptocephala walcotti, Redlichia rex, Olenellus gilberti, Mesonacis fremonti, Gabriellus kierorum, Lochmanolenellus trapezoidalis, Fallotaspis longa, Peachella iddingsi, Paradoxides gracilis, Zhangshania typica

ODONTOPLEURIDA
(LOWER ORDOVICIAN—UPPER DEVONIAN)

Dicranurus hamatus elegantus, Chlustinia keyserlingi, Boedaspis ensifer, Odontopleura markhami, Kettneraspis williamsi, Miraspis mira, Leonaspis deflexa, Meadowtownella trentonensis, Acidaspis cincinnatiensis, Ceratocephala barrandei

TRINUCLEIDA
(MIDDLE CAMBRIAN—UPPER SILURIAN)

Onnia superba, Paratrinucleaus acervulosus, Declovithus titan, Lloydolithus lloydi, Tretaspis sortita, Trinucleus fimbriatus, Stapeleyella inconstans, Marrolithus ornatus, Tretaspis limbatus, Nankinolithus granulatus

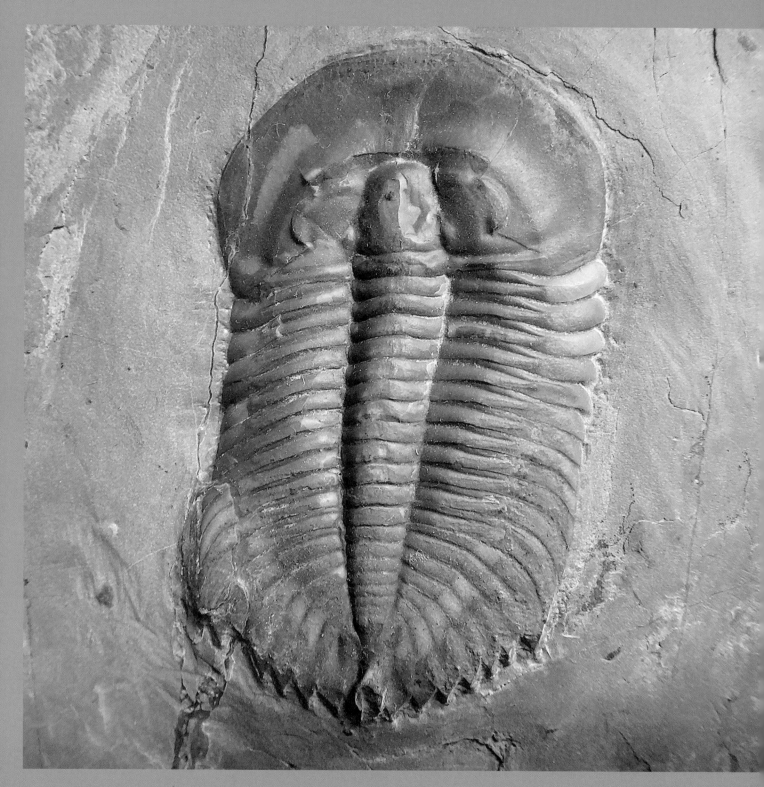

MONKASPIS DAULIS (WALCOTT, 1905)

Middle Cambrian; Damesella paronai Zone, Kushan Formation; Shandong Province, China; 9.1 cm

ORDER: Asaphida

(LEFT) *XIPHOGONIUM TRAUTENSTEINENSIS CRASSUS* (G. K. B. ALBERTI, 1969)

Lower Devonian; Timrhanrhart Formation; Boutshrafine, Morocco; 2.3 cm

ORDER: Proetida

(BOTTOM) *NANKINOLITHUS CF. GRANULATUUS* (WHALENBERG, 1818)

Upper Ordovician, Ashgill Stage, highest Rawtheyan Substage; Upper Drummuck Group, Farden Member; southeastern (highest) end of "Rae's Trench"; Lady Burn, Girvan, Ayrshire, Scotland; 3.4 cm

ORDER: Trinucleida

PARACERAURUS SP.

Middle Ordovician, Lower Llandeilian; *Asaphus kowalewski* **Layers; St. Petersburg, Russia; 12.8 cm**

ORDER: Phacopida

CERAURUS MATRANSERIS SINCLAIR, 1947

pper Ordovician; Lindsay Formation; Colborne, Ontario, Canada; 6.6 cm

RDER: Phacopida

POROSCUTELLUM SP.

Lower Devonian, Pragian Stage; Ihandar Formation; Tafraoute, Morocco; 4 cm

ORDER: Corynexochida

45

10 SIGNIFICANT
SOFT-TISSUE TRILOBITES

Ever since the first fossil discoveries were made over a century ago in the Middle Cambrian layers of British Columbia's famed Burgess Shale, the paleontological community has been acutely aware that underneath their hard calcite exoskeletons trilobite anatomy also featured an impressive array of nonbiomineralized limbs and organs. These included antennae, gills, walking legs, and even basic respiratory and digestive systems, all of which have been effectively and in some cases magnificently preserved in various Paleozoic horizons around the globe. Even earlier, in the Lower Cambrian when oceans dominated the planet and surrounded the supercontinents Laurentia and Amazonia, trilobite species such as *Olenellus getzi* and *Kuanyangia pustulosa* (both have displayed fossilized evidence of legs and antennae) were joined in their marine ecosystem by a plethora of other invertebrate life-forms. Many of these were soft-tissue arthropods whose existence has been chronicled in paleontological treasure troves found in Morocco, Australia, and the western United States. Perhaps even more intriguing, in southern China, among the Lower Cambrian Chengjiang Biota's stunning variety of soft-body fossils, there is evidence of the first chordates. These include *Yunnanozoon lividum*, a tiny animal presenting what appears to be a primitive backbone. Soft-part preservation is certainly not relegated solely to Earth's Cambrian strata. Recent Ordovician discoveries in North Africa's Fezouata Formation have revealed previously unseen and unknown aspects of trilobite anatomy, including digestive tracts, and similarly aged finds in the Beecher's Trilobite Bed deposits of upstate New York

ANACHEIRURUS SP.

Lower Ordovician, Florian Stage; Fezouata Formation; Zagora region, Morocco; trilobite with antennae 5.3 cm

have provided even more dramatic insight into trilobite soft-body morphology, showing clear signs of both eggs and primitive but apparently highly efficient reproductive systems.

Here's a look at 10 trilobites that exhibit significant soft-tissue preservation.

ANACHEIRURUS SP.

Few recent paleontological finds are more significant than the 1998 discovery of Morocco's Fezouata Formation. In the heart of the Sahara Desert, layer upon layer of fossil-filled Lower Ordovician strata has been revealed—some showcasing bizarre soft-bodied creatures that exhibit unmistakable affinities to Middle Cambrian arthropods unearthed in other parts of the planet. The academic community quickly jumped on the Fezouata's revelatory and visually stunning specimens, as did collectors from Casablanca to Cucamonga. Although the formation's nonbiomineralized examples display striking similarities to fauna found in outcrops such as the legendary 30-million-year-older Burgess Shale, the trilobites—like the *Anacheirurus* specimen shown here displaying antennae—have generally proved to be indigenous only to this prime Paleozoic outcrop.

WUTINGASPIS TINGI

The now-legendary Chengjiang Biota is located in the heart of Southern China's bustling yet often bucolic Yunnan Province. Since its scientific discovery in the mid-1980s, the site has become world renowned for its outstanding faunal diversity and for its unique preservation of a wide assortment of nonbiomineralized life-forms. These 518-million-year-old specimens include four species of trilobites, including *Wutingaspis tingi*. When carefully prepared, these mid-sized trilobites, that have been found up to 11 centimeters in length, may show evidence of antennae and (even less frequently) walking legs.

OLENELLUS GETZI

During the middle and late decades of the twentieth century, the famed Lower Cambrian quarries of eastern Pennsylvania were among the most heavily worked fossil zones in the world. Hardly a day would go by without teams of enthusiastic amateurs from neighboring communities flocking to the region and attempting to break apart the hard Kinzers Formation limestone layers in search of prized *Olenellus* and *Wanneria* specimens. Among the hundreds of complete trilobites recovered, only a scant few would later reveal evidence of soft-tissue preservation. Many of these *Olenellus getzi* examples are now housed in Yale's Peabody Museum.

TRIARTHRUS EATONI

The startling soft-tissue preservation captured on the 440-million-year-old *Triarthrus* specimens found in upstate New York's Lorraine Shale has been noted and studied for more than a century. But only in recent years has contemporary research (along with improved preparation techniques) begun to reveal the amazing detail of trilobite internal anatomy available for observation in these uniquely pyritized specimens. During the past decade, legs, gills, antennae, and even eggs have been noted amid the hundreds of complete examples uncovered at the famed Ordovician-age Beecher's Trilobite Bed site.

MEGISTASPIS HAMMONDI

In 2016, fossils unearthed in the 478-million-year-old Fezouata Formation of Morocco supplied key information about what trilobite internal body design—and subsequent behavior—may have been like in those long distant Ordovician seas. Working with three exceptionally well-preserved examples of the unusual, long-tailed asaphid *Megistaspis hammondi*, a pair of Spanish scientists uncovered evidence of a previously unseen midgut gland, which in many modern arthropods is known to secrete enzymes used for the digestion of food. Their studies also exposed signs of a crop, an internal organ commonly found in contemporary sediment feeders, especially those that live along the ocean floor. Other examples of this species have revealed prominent, strangely segmented antennae.

CHOTECOPS FERDINANDI

In some special cases, careful preparation may reveal tantalizing glimpses of a trilobite's soft-tissue morphology—if, of course, the specimen in question managed to retain evidence of its appendages during the fossilization process. At other times, those hidden legs, antennae, and gills are best revealed when subjected to high-power X-ray technology. Such is the case with certain examples of the phacopid genus *Chotecops* found in Germany's Devonian-age Hunsruck Shale. These relatively common 6- to 9-centimeter-long trilobites often only reveal their complex soft-part anatomy after the slate sheets in which they've been fossilized are zapped by a bit of radiant energy.

EOREDLICHIA INTERMEDIA

Trilobites bearing nonbiomineralized limbs and appendages were first revealed in China's Lower Cambrian Chengjiang Biota late in the twentieth century. At that time, both the academic and collecting communities were pleasantly surprised by the sheer volume of material that featured such an unusual and highly coveted state of preservation. Although far from common, trilobites such as *Eoredlichia intermedia* were apparently rather routinely fossilizing in a manner that preserved their soft-tissue body parts—especially antennae—in that formation's fine-grained limestone layers.

OLENOIDES SERRATUS

Lurking amid the more acclaimed army of soft-bodied arthropods that line the Burgess Shale's renowned Middle Cambrian outcrops are an impressive variety of trilobite species, including *Olenoides serratus*. Both dorsal and ventral examples of *Olenoides* have been found with well-preserved walking legs and antennae prominently featured in their fossilized forms. How often these specimens were subjected to this distinctive manner of preservation remains unclear, but for well over a century both the scientific and collecting fraternities have celebrated the fact that these trilobites have occasionally revealed the secrets of their soft-part anatomy.

AMPHOTON DEOIS

Since the dawn of the twenty-first century, perhaps no nation on Earth has been the source of more diverse trilobite discoveries than China. From the countless poorly preserved Cambrian and Ordovician examples that continually pop up on eBay to the magnificent specimens that now frequently highlight international fossil extravaganzas, this vast Asian landmass has emerged as one of the planet's most notable Paleozoic repositories. A key member of that extensive Chinese trilobite fauna is the diminutive Middle Cambrian species *Amphoton deois* (usually no more than 2 centimeters in length), which is perhaps most renowned for the occasional preservation of its gut tract. These can appear as dark blotches in either the trilobite's cephalic or thoracic regions.

REDLICHIA TAKOOENSIS

Since the discovery of Australia's Emu Bay Shale in 1955, trilobites and soft-bodied arthropods of all sorts and sizes (including fragments of the legendary Terror of the Cambrian Seas, *Anomalocaris*) have been unearthed. The coarse, reddish-brown mudstone that comprises this Lower Cambrian deposit does not preserve its fossil fauna with the same degree of detail enjoyed by specimens found in the Early Paleozoic Lagerstatte of western Canada and southern China. However, on rare occasions, Emu Bay examples of the large trilobite species *Redlichia takooensis* have been uncovered with evidence of their antennae still intact.

(OPPOSITE PAGE) **WUTINGASPIS TINGI (KOBAYASHI, 1944)**

Lower Cambrian; Chengjiang Maotianshan Shales, Quiongzhusi Section; Heilinpu Formation; Yunnan Province, China; 5.4 cm

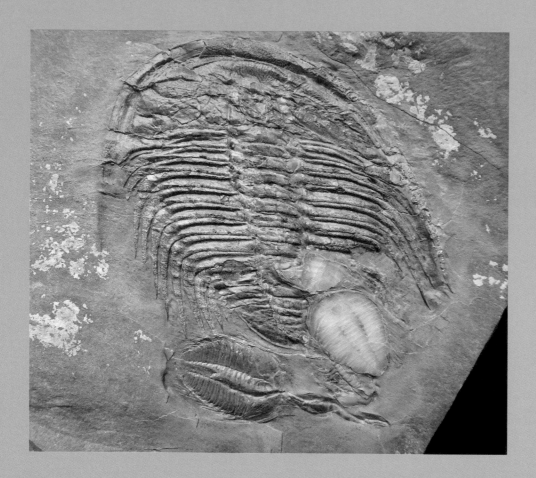

(TOP) *ORYGMASPIS (PARABOLINOIDES) MCKELLARI CHATTERTON AND GIBB, 2016*

Upper Cambrian, Furongian; McKay Group, Taenicephalus Zone; Cranbrook, British Columbia, Canada; 6.4 cm

Note calcite trilobite on bottom right.

(RIGHT) *TRIARTHRUS EATONI (HALL, 1838)*

Upper Ordovician; Lorraine Shale, Martin Quarry; Beecher's Trilobite Bed; Oneida County, New York, United States; 0.9 cm

Photo courtesy of Markus Martin

MEGISTASPIS (EKERASPIS) HAMMONDI CORBACHO AND VELA, 2010

Lower Ordovician, Arenigian; Upper Fezouata Shales; Draa Valley, Zagora, Morocco; 18.7 cm with antennae

AMPHOTON DEOIS (WALCOTT, 1905)

Middle Cambrian, Series 3, Drumian; Zhangxia Formation; Liaoning, Shandong Province, China; 2.2 cm

Note possible preservation of internal organs.

46

10 TOP TRILOBITE MUSEUMS

A visit to any natural history museum is an enlightening experience, whether that institution is located in the heart of a major North American city or on the outskirts of a nondescript European village. Packed with artifacts, antiquities, and fossils from all corners of the globe and all aspects of time, for everyone from grade-school kids to grizzled college professors, these museums are among the world's most engaging places. Indeed, many of these venues serve as the premier tourist destination in their respective town or country, repositories capable of attracting hundreds of thousands—if not millions—of visitors annually, all eager to enjoy that establishment's featured exhibits, lifelike dioramas, and enlightening paleontological displays. Yet even in the most prestigious natural history museums, trilobites rarely receive their proper degree of recognition and respect. Too often the fossilized remains of these ancient organisms are relegated to small cases in dark corners of back rooms, while the oversized bones of dinosaurs and woolly mammoths vie for the saber cat's share of front lobby acclaim. Despite the less than stellar manner in which trilobites are frequently presented to the public, a number of notable institutions around the globe have at least attempted to prioritize their collection and display of trilobites.

Here is a brief look at 10 of the world's top trilobite museums.

AUSTRALIAN MUSEUM, SYDNEY

Located in the middle of Sydney's bustling downtown business district, the Australian Museum has amassed extensive collections over the last two centuries that run the gamut from indigenous birds to Aboriginal artifacts. The museum also houses the world's foremost assemblage of Lower Cambrian fossils drawn from the Emu Bay Shale of Kangaroo Island, some 1,500 kilometers to the southwest. Those holdings include an impressive array of that formation's most formidable trilobite species, *Redlichia takooensis* and *Redlichia rex*—some up to 25 centimeters in length and occasionally bearing predatory bite marks—as well as a revealing selection of soft-bodied material that is comparable to but considerably older than the more famous nonbiomineralized fossils drawn from the Burgess Shale of British Columbia.

Australian Museum: 1 William Street, Darlinghurst, NSW, Australia

ROYAL ONTARIO MUSEUM, TORONTO

The Royal Ontario Museum conveniently houses the planet's largest collection of Burgess Shale material, so it should come as no great surprise that an eye-popping display of that BC locale's hallowed Middle Cambrian fauna—featuring a formidable assembly of soft-bodied organisms—is on exhibit in this all-encompassing Canadian facility. The museum's recent and impressively updated Paleozoic presentation also includes such unique trilobites as an *Olenoides serratus*, with soft-tissue preservation, and a virtually complete *Terataspis grandis* specimen found less than 200 kilometers from the museum's front door. Of particular note is a case featuring perhaps the only semicomplete example of the famed "trilobite eater" *Anomalocaris canadensis*.

Royal Ontario Museum: 100 Queens Park, Toronto, Canada

AROUCA GLOBAL GEOPARK, PORTUGAL

In the early years of the twenty-first century, workers extracting thick sheets of Ordovician-age slate from the Canelas Quarry in Arouca, Portugal, stumbled on something extraordinary—a layer of huge, flattened trilobites up to 70 centimeters in length that in some places virtually blanketed the surrounding rock surface. Featuring species such as *Ogyginus forteyi* and *Ectillaenus giganteus*, these ghostly white trilobites caused an immediate stir in both academic and collecting communities. Now designated as a UNESCO Global Geopark, which both celebrates the location's amazing discoveries and protects these giant trilobites for future generations, the town of Arouca has recently built a small but significant on-site museum that presents some of the quarry's major finds and also highlights the history and scientific importance of the 450-million-year-old specimens revealed in its charcoal-gray strata.

Global Geopark Museum: R. Alfredo Van Pinto, 4540–118 Arouca, Portugal

AMERICAN MUSEUM OF NATURAL HISTORY, NEW YORK

A convenient stroll from just about anywhere on the island of Manhattan, this legendary Big Apple institution is certainly most renowned for its grand dinosaur galleries that encircle the entire fourth floor of the museum. A small but comprehensive trilobite display is located in the AMNH's 77th Street lobby. Among the baker's dozen of featured specimens—suitably presented in proper time-sequence order—are a pair of spectacular Silurian offerings: *Spathacalymene nasuta* from Indiana and

PROCERATOPYGE SP.
Upper Cambrian; Huayansi Formation; Huayuan County, Jishou City, Hunan, China; 4.7 cm

***PSEUDOSPHAEREXOCHUS (PATERASPIS) YINGANENSIS* ZHANG TAIRONG, 1981**
Upper Ordovician, Caradoc—Early Ashgill; Shihtien Series, Pupiao Formation; Shidian, Baoshan, Yunnan Province, China; 5.1 cm

Arctinurus boltoni from New York. There is also a massive *Xenasaphus devexus* plate from the Ordovician outcrops that surround St. Petersburg, Russia.

American Museum of Natural History: 200 Central Park West, New York City, New York

NATURAL HISTORY MUSEUM, LONDON

Known until 1992 as the British Museum (Natural History), it is now recognized simply as the Natural History Museum. This imposing structure in central London houses a splendid assemblage of indigenous trilobites, some of which adorn a series of sturdy, prime-location display cases. There is a definite (and understandable) emphasis placed on the magnificently preserved Silurian specimens from nearby Dudley, but among the other trilobites cached behind this institution's dramatic Romanesque facade (first opened in 1881) are an intriguing selection of *Ogygiocarella* examples hailing from throughout the British Isles.

SMITHSONIAN INSTITUTE, WASHINGTON, D.C.

Recently renovated, the Paleozoic exhibits presented in the National Museum of Natural History (more commonly known to one and all as the Smithsonian) feature some of the most impressive trilobite specimens to be found anywhere in the world. Among the key items on display are a rare *Bathynotus holopygus* from Vermont and an equally unusual *Trimerus vanuxemi* from West Virginia. The only known complete *Apianurus sp.* from the hallowed Walcott/Rust Quarry of New York State is also showcased there. However, as nice as the new exhibits are, many longtime visitors still miss the old trilobite cases that included a large *Isotelus maximus* from Ohio and a complete, 12-centimeter *Dikelocephalus minnesotensis* from Wisconsin.

Smithsonian Institute: 10th Street and Constitution Avenue NW, Washington, D.C.

HUNTERIAN MUSEUM AND ART GALLERY, GLASGOW

The beautifully preserved Ordovician trilobites of Girvan, Scotland, have long been collected, admired, and studied around the globe. And no museum contains a more comprehensive grouping of this distinctive Paleozoic material than the University of Glasgow's own Hunterian Museum. Enriched over the decades by donations from many notable local collectors—some of whom began their digging adventures in the early years of the nineteenth century—this institution both houses and displays a rich variety of the area's outstanding fossil fauna, including stellar examples of such trilobites as *Paracybeloides girvanensis* and *Uripes maccullochi.*

HOUSTON MUSEUM OF NATURAL SCIENCE, TEXAS

This relatively new, expansive, and seemingly still growing facility in the American southwest highlights what is perhaps the continent's largest and most impressive public trilobite display. Inside the museum, more than 150 superlative examples, drawn from across the face of the planet and representing every Paleozoic period, can be viewed in a sequence of spacious, well-lit cases. This impressive assemblage is primarily focused on the aesthetic presentation of exotic Moroccan and Russian species, most originally obtained from the collection of noted Houston-based trilobite authority Sam Stubbs. A choice number of North American trilobites are also showcased, including *Elliptocephala sp.* from Nevada and *Gabriceraurus dentatus* from Ontario.

Houston Museum of Natural Science: 5555 Hermann Park Drive, Houston, Texas

CZECH NATIONAL MUSEUM, PRAGUE

The Czech National Museum houses a preponderance of the legendary Bohemian Jince Formation material gathered in the nineteenth century by the famed Joachim Barrande. Prior to its recent (and seemingly never-ending) renovation, this magnificent facility in the center of Prague presented one of the most extensive and comprehensive regional trilobite displays to be found in Europe. Fewer specimens are on exhibit now, but they are under better lighting conditions and without decades-old dust marring the view of any interested observer. The museum highlights include breathtaking examples of locally discovered *Bumastus hornyi* and *Dalmanitina socialis*, photos of which have long graced the pages of the planet's leading scientific journals.

Czech National Museum: Vaclavske nam. 68, Nove Mesto, Prague, Czech Republic

MUSEUM OF COMPARATIVE ZOOLOGY, CAMBRIDGE, MASSACHUSETTS

Visitors hoping to be overwhelmed by the myriad trilobite specimens housed in this medium-sized facility located on the campus of Harvard University will invariably be disappointed. Indeed, only a handful of so-so examples are on public display. But in its well-maintained back rooms, the Museum of Comparative Zoology possesses one of the premier trilobite collections in the world. In addition to an outstanding arthropod array culled from New York's Walcott/Rust Quarry (many acquired directly from Charles Walcott himself), the museum features the most complete European trilobite assemblage to be found in a North American institution—including an extensive Barrande-sourced hoard of Czech material—as well as a unique gathering of the locally discovered *Acadoparadoxides harlani* from nearby Braintree, Massachusetts.

Museum of Comparative Zoology: 26 Oxford Street, Cambridge, Massachusetts

(OPPOSITE PAGE) **NOBILASAPHUS SP.**
Lower Ordovician; Draa Valley, Zagora area; Lower Fezouata Formation, Morocco; 16.5 cm

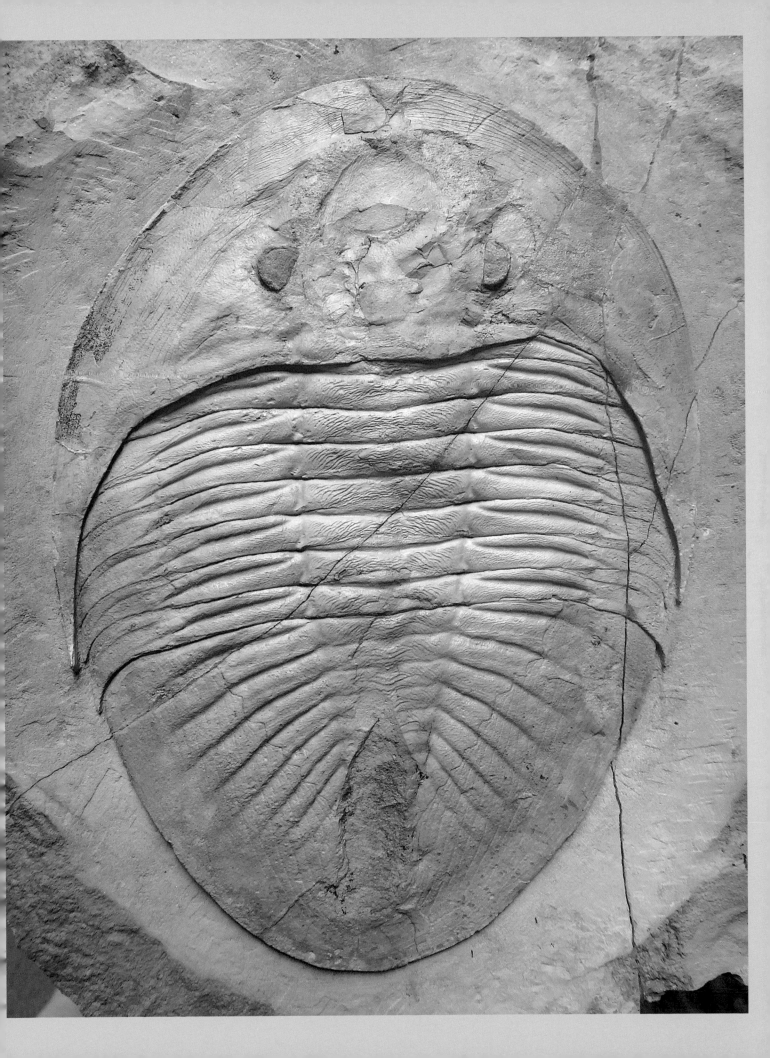

(TOP) *DIKELOKEPHALINA SP.*

Lower Ordovician, Arenigian; Upper Fezouata Shales; Draa Valley, Zagora, Morocco; 8.4 cm

(LEFT) *HEMIRHODON AMPLIPYGE* ROBISON, 1964

Middle Cambrian; Marjum Formation; Millard County, Utah, United States; 8.8 cm

(OPPOSITE PAGE) *SUBASAPHUS PLATYURUS LATISEGMENTATUS* (NIESZKOWSKI, 1857)

Middle Ordovician, Upper Llanvirnian; Aseri Regional Stage, Lower *Asaphus kowalewskii* Zone; Duboviki Formation; Vilpovitsy Quarry, St. Petersburg region, Russia; 7.8 cm

ELLIPTOCEPHALA SP.

Lower Cambrian, Series 2, Lower Dyeran; Poleta Formation; Montezuma Mountains, Nevada, United States; 7.3 cm

10 TOP TRILOBITE PRETENDERS

47

Following their demise at the end of the Permian Period some quarter-of-a-billion years ago, it is an irrefutable scientific fact that the entire trilobite class came to an abrupt and eternal end. Quite simply, after their incredible journey through more than 270 million years of evolutionary time, these highly resilient invertebrates vanished, leaving no living descendants behind. A variety of contemporary arthropods may share a smattering of morphological features with trilobites—including legs (first noted in trilobite fossils in 1870), an easily molted external shell, the placement of their eyes, and even a multijointed thorax—but other equally important anatomical aspects of their exoskeletons, such as eye ridges and the shapes of their hypostome, mark trilobites as a totally separate branch of the arthropod family tree. So why did trilobites leave nary a single descendant in their Paleozoic wake? The primary reason academics cite for the lack of modern trilobite offshoots is that from the moment of their emergence more than 521 million years ago they were the most archaic members of the arthropod line. For example, the first chelicerates (i.e., spiders) and crustaceans (i.e., shrimp) didn't appear on the marine scene for another 10 to 15 million years. Another rationale for the trilobite line's descendant-free demise is that throughout their Paleozoic passage these unique organisms found themselves serving an increasingly less significant role in the water world surrounding them. Thousands of trilobite species dominated the Cambrian seas, but by the conclusion of their crawl through time in the Permian, their size, numbers, and dominance had been reduced to mere blips on the evolutionary scorecard.

Here's some insight on 10 modern-day—along with a few fossilized—trilobite pretenders.

ISOPODS

At least once a month, a visitor to one of the seemingly countless trilobite-related Facebook or Instagram pages will post a photo of an isopod accompanied by some sort of "living trilobite?" caption. Ummm . . . somewhat intriguing, but no. Although there unquestionably is a strong superficial resemblance between these contemporary crustaceans and certain members of the long gone trilobite line (including a hard segmented exoskeleton and numerous walking legs), isopods are certainly not trilobite descendants—even though one notably mimicking species is named *Serolis trilobitoides*.

HORSESHOE CRABS

No one interested in trilobites—and in possession of an even mildly vivid imagination—has ever seen a horseshoe crab carapace washed up on some distant shoreline and not imagined a somewhat similar trilobite-themed Deep Time scenario. These members of the genus *Limulus* are perhaps the closest living trilobite relative, but in no way, shape, or fossilized form are they a direct descendant of the noble trilobite class. Aside from their common arthropod ancestry, both also may have shared a similar lifestyle characteristic—swimming upside down.

SPIDERS

In the 1950s, "B movie" sci-fi flicks thrived on presenting large fake spiders to their viewing audience, occasionally trying to pass these creatures off as some sort of mutant, A-bomb-infected trilobite replicant. Quite obviously, they weren't! There is, in fact, a lot of taxonomic territory separating the trilobite and chelicerate lines. As a handy point of reference, the arthropod phylum is divided into four basic subgroups. These include hexapods (insects such as ants and bees), myriapods (centipedes and millipedes), crustaceans (shrimp and crabs), and chelicerates (spiders, scorpions, and horseshoe crabs). Trilobites are related to all but are direct ancestors to none.

CHITONS

Many of us have walked along a rock-strewn beach jetty on a warm summer's day and seen chiton shells firmly attached to the adjacent stone surfaces. These midsized mollusks feature a segmented, plated shell that on first impression is somewhat reminiscent of a fossilized trilobite carapace. But when pried from their rock-strewn crevices, the chiton underside reveals a prominent snail-like "foot" rather than the hoped for trilobite-clone walking legs.

EURYPTERIDS

Few intimately familiar with the fossil world have ever confused these (mostly) Silurian arthropods with trilobites, but they are include on this somewhat expansive "pretenders" list because they shared some of the same Paleozoic seas. However, trilobites were believed to be strictly salt water inhabitants, whereas eurypterids apparently preferred brackish waters. It has been speculated that large eurypterids occasionally may have feasted on trilobites, although no fossil evidence supporting this supposition has yet been uncovered.

TADPOLE SHRIMP

With their broad head shields, prominent eyes, notable antennae, and sturdy rows of walking legs,

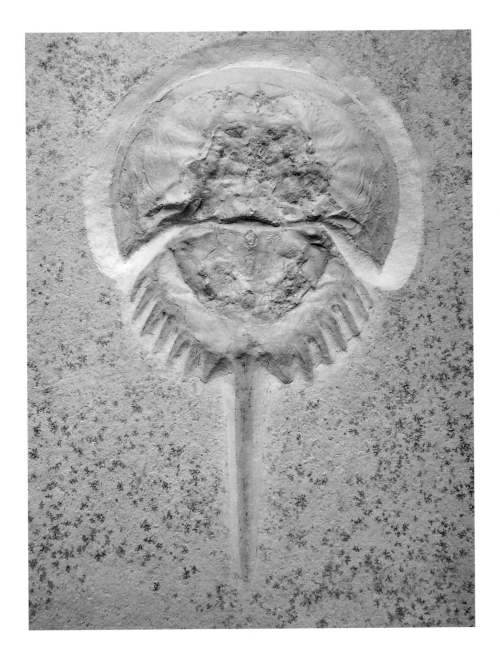

MESOLIMULUS WALCHI
STORMER, 1952

**Upper Jurassic, Kimmeridgian Stage;
Solenhofen Limestone; Eichstatt,
Germany; 9.7 cm**

An ancestral horseshoe crab.

it's easy to understand why those not intimately familiar with trilobite morphology may initially mistake tadpole shrimp for "living fossils," if not exactly for living trilobites. However, the arbitrary and often whimsical elements of parallel evolution exhibited by these shallow-water crustaceans are perhaps more a reflection of how advanced and efficient certain aspects of the trilobite anatomy were, even half-a-billion years ago.

EUPROOPS

Although their fossilized remains are found worldwide, the ancient xiphosuran known as *Euproops* is a particular favorite for many who collect the famed Carboniferous-aged Mazon Creek fauna of the midwestern United States. Closely related to modern-day horseshoe crabs, many collectors who've stumbled on the positive/negative splits of these distinctive arthropods have wondered if

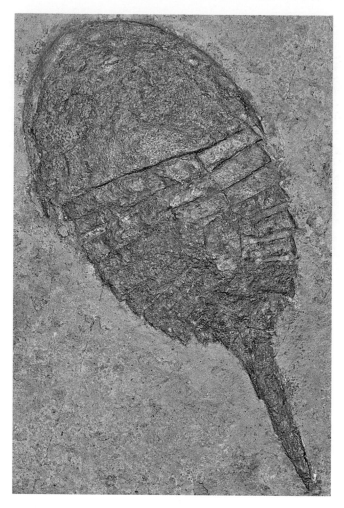

***BECKWITHIA TYPA* RESSER, 1931**

Middle Cambrian; Weeks Formation; House Range, Utah, United States; 20.3 cm

These large aglaspid arthropods lived alongside trilobites.

they may have unearthed some sort of post-Paleozoic trilobite, or at least a trilobite offshoot. Rest assured, *Euproops* is neither.

WATER PENNIES

It may be stretching things a bit to claim that the contemporary fresh-water creatures commonly known as Water Pennies have anything at all to do with trilobites—past, present, or future. These organisms merely represent the larval stage of the *Mataeopsephus* genus of beetle—a fact that clearly marks them as insects rather than trilobite-like arthropods. Yet those who encounter these hard-shelled organisms can be momentarily fooled into believing they have stumbled on a momentous scientific discovery.

XANDARELLIDS

The fossilized remains of strange, segmented arthropods known as *Phytophilaspis pergamena*—a representative of the xandarellid evolutionary line—lurk in Siberia's Lower Cambrian Sinsk Formation. These large 10- to 15-centimeter-long animals bear a striking superficial resemblance to some early trilobites and share such characteristics as prominent eyes, a large pygidium, and even the outline of their hypostome. But they are also quite different due to the shape of their facial sutures, the reduced size of their thorax, and the fact that their limited number of thoracic pleura are fused to one another.

LOBSTERS

The running (or should we say crawling) joke among certain trilobite aficionados is that they may never know what those ancient arthropods tasted like, but they certainly hope it was something akin to their modern-day arthropod relative, the lobster. Lobsters have also provided the academic community with vital information regarding arthropod size variance and longevity (they can weigh up to 20 kilograms and live up to 100 years) that in imaginative hands may be applied to certain forms of trilobite research.

EURYPTERUS REMIPES DEKAY, 1825

Upper Silurian; Bertie Group; Fiddler's Green Formation; Herkimer County, New York, United States; larger specimen 17.3 cm

These chelicerates shared the Silurian seas with trilobites.

PHITOPHILASPIS PERGAMENA IVANTSOV, 1999

Lower Cambrian; Botoman Regional Stage; Sinsk Formation; Botoma River (right tributary of Lena River); Lena Pillars, Southern Yakutia, Sakha Republic, Eastern Siberia; 12 cm

An early arthropod related to trilobites.

48

10 WONDERFUL WALCOTT/RUST TRILOBITES

New York State's Walcott/Rust Quarry enjoys a long and noble history. Initially opened for fossil exploration by the landowner William Rust in the middle of the nineteenth century, by 1870 he had been joined in his efforts by 20-year-old Charles Walcott, then four decades before his historic discoveries at British Columbia's Burgess Shale. More than a century later, the noted trilobite enthusiast Thomas Whiteley brought a new level of academic renown to this revered Ordovician-age quarry. His efforts were closely followed by a commercial venture launched by the father-son team of Dan and Jason Cooper. Eschewing some of the scientifically inclined subtlety Whiteley had employed during his efforts, in 2008 the Coopers began the most extensive excavation the quarry had yet experienced. Their goals were clear: to find, prepare, and then market the best material ever uncovered at the famed site. They believed the specimens thus revealed would rank in quality with trilobites found in any other fossil-bearing outcrop on Earth. With their thick black shells, intricate surface ornamentation, and incredible three-dimensional preservation, the 18 distinct trilobite species pulled from these fauna-rich 450-million-year-old layers proved to be spectacular. Some examples were so lifelike in their fossilized form that it seemed all one needed to do was drop them in water and watch them crawl away. There were huge football-shaped examples of *Isotelus gigas*, along with pristine displays of *Flexicalymene senaria* and *Meadowtownella trentonensis*. Especially rare species, such as the delicately spined 8-centimeter-long *Apianurus sp.*, went directly into exhibits at the Smithsonian. Other

AMPHILICHAS CORNUTUS (CLARKE, 1894)

Middle Ordovician; Rust Formation, Trenton Group; Walcott/Rust Quarry; Trenton Falls, New York, United States; 6.2 cm

unusual trilo-types, such as the diminutive bubble-nosed *Sphaerocoryphe robusta* and the elegant *Amphilichas cornutus*, ended up in private collections everywhere from Barcelona, to New York City, to Cancun. These latest arthropod-inspired adventures have added an exciting new chapter to the Walcott/Rust Quarry's already storied tale.

Here's a salute to 10 wonderful Walcott/Rust trilobites.

AMPHILICHAS CORNUTUS

Trilobite enthusiasts are an unusual group, and most would be first in line to confess their fossil-fueled foibles. Like many collectors around the globe—whether their focus is on baseball cards, Barbie dolls, or bottle caps—trilobite collectors often want nothing more than what they can't easily possess, and certainly near the top of any such "want" list would be a prime example of the super rare species *Amphilichas cornutus*. Despite decades of digging at the Walcott/Rust site, only three complete specimens of this pustule-covered trilo-type have been found. As recently as 2020, however, two additional partial pieces (both, in Anne Boleyn fashion, missing their heads) had been recovered, reviving hope that more samples of this beautiful trilobite will appear in the short-term future.

CERAURUS PLEUREXANTHEMUS

After decades of disregard and decay, by late in the twentieth century the mere existence of the Walcott/Rust Quarry site had become little more than a rumor to many in the fossil field. Despite the elegant beauty and scientific importance of trilobites such as *Ceraurus pleurexanthemus*—the first documented trilobite species to display fossilized evidence of legs—and the closely related but exceptionally rare *Ceraurus sp.*, most in the

CERAURUS SP.
Middle Ordovician; Rust Formation, Trenton Group; Walcott/Rust Quarry; Trenton Falls, New York, United States; 4.1 cm

Photo courtesy of the Martin Shugar Collection

academic community showed only a passing interest in this long-lost locality in central New York State. Those who chose to focus their attentions on the region's magnificently preserved Ordovician trilobites invariably contented themselves with studying the impressive array of material already housed in the hallowed halls of universities such as Harvard and Yale.

ISOTELUS GIGAS

These large, football-shaped trilobites—on occasion they attained dimensions rivaling the size of those sporting spheres—can be found in multiple assemblages that represent some of the most impressive and expensive fossil plates in the world. It is believed that one noted Asian collector recently spent nearly $60,000 on a large Walcott/Rust fossil block featuring six 15-centimeter-long perfectly preserved *Isotelus gigas* specimens. These imposing morphologically streamlined trilobites often appear in fossilized form alongside a variety of faunal elements—including other trilobite species—a fact that provides invaluable insight into possible undersea interactions along the Ordovician ocean floor.

FLEXICALYMENE SENARIA

Members of the *Calymene* genus fill Ordovician Period fossil formations in localities across the face of the planet. Whether it's in Ohio, England, Morocco, or eastern Canada, various species of this ubiquitous trilobite are as familiar to most collectors as their favorite pair of well-worn shoes. *Flexicalymene senaria* is certainly no exception to this Paleozoic standard. Although perhaps not as common a Walcott/Rust faunal component as *Ceraurus pleurexanthemus*—the quarry's most prolific trilobite—enough of these attractive, black-shelled bugs have been unearthed to allow them to be compared and contrasted to similar Flexi examples found throughout the world.

MEADOWTOWNELLA TRENTONENSIS

It's not exactly reinventing the Paleozoic wheel (or expanding on the fundamental concepts of, say, punctuated equilibria) to state that big trilobites are generally easier to find than little ones. Such fundamental logic is certainly in play within the revered walls of the Walcott/Rust Quarry where 15-centimeter-long Ordovician behemoths such as *Isotelus gigas* are routinely found. Small (generally 1- to 2-centimeter-long) species such as *Meadowtownella trentonensis*—or the equally diminutive

and even less common *Diacanthaspis parvula*—are among the rarest of recovered examples. But searching for these tiny trilobite treasures is certainly worth the effort, especially if you're fortunate to possess 20/20 eyesight!

BUMASTOIDES HOLEI

According to various scientific papers that have chosen to address this trilo-centric subject, two recognized species of *Bumastoides* appear in the hard limestone layers of the Walcott/Rust Quarry: *B. holei* and *B. porrectus*. Perhaps it is properly left to the cerebral members of the academic community to best explain the differences between these two; from a layman's perspective let's just say that they are minimal. Considering that many of those early-twentieth-century taxonomic decisions were based on partial or poorly preserved examples, perhaps a slight revision of the Walcott/Rust *Bumastoides* fauna is well past due.

HYPODICRANOTUS STRIATULUS

Looking like a misplaced visitor from an alien world, few nonspiny trilobite species have ever drawn more mainstream "oohs and aahs" than the tongue-trippingly named and hydrodynamically shaped *Hypodicranotus striatulus*. Due to the scarcity of complete specimens—with less than half-a-dozen articulated examples of *Hypodicranotus* emerging from the Walcott/Rust layers despite decades of on-site digging—the most recognized morphological feature of this unusual, 3-centimeter-long species is its elongated hypostome, or mouth plate. These forklike projections, which ran almost the complete ventral length of the creature's body, are occasionally found independently, providing tantalizing clues about this species' presence in the quarry's nearly half-billion-year-old sedimentary outcrops.

CERAURUS PLEUREXANTHEMUS (GREEN, 1832)
Upper Ordovician; Bobcaygeon Formation; Belleville, Ontario, Canada; 3.2 cm

The same species is abundant within the W/R Quarry.

SPHAEROCORYPHE ROBUSTA

When the ongoing commercial dig at the Walcott/Rust Quarry commenced in 2008, one of the most surprising and exciting discoveries was of a layer featuring well-preserved examples of the small, bulbous-nosed cheirurid *Sphaerocoryphe robusta*. Disarticulated fragments of this 450-million-year-old trilobite have been recovered dating all the way back to the site's original nineteenth-century digs, but complete specimens of the compact *Sphaerocoryphe*—often no larger than a single centimeter—were always exceptionally rare commodities.

Fewer than a score have been revealed during recent excavations, and those have quickly been gobbled up by eager collectors and museum officials around the globe.

ACHATELLA ACHATES

The planet-spanning Ordovician genus *Achatella* was first widely recognized from examples discovered during mid-nineteenth-century digs conducted in Great Britain and Russia. It is also a trilobite known from a variety of midwestern locations, including outposts in Indiana and Missouri. But the specimens discovered in the Walcott/Rust Quarry hold a unique collector cachet due to both their lustrous charcoal-gray calcite preservation and their often perfectly aligned, lifelike appearance. With detailed compound eyes perched high atop their gracefully curving cephalons, along with their streamlined body shape, examples of *Achatella achates* rank among the more attractive and compelling of all Walcott/Rust faunal components.

ISOTELUS WALCOTTI

Science-based debates can be truly fascinating, especially when both sides of these highbrow discussions believe strongly in the case they are presenting. With that in mind, members of the trilobite collecting community insist that *Isotelus walcotti* must be considered among the rarest elements of the Walcott/Rust Quarry's indigenous fauna. Others state with equal fervor that these relatively small, 4- to 7-centimeter-long specimens—all of which feature pronounced genal spines—are nothing more than immature examples of the omnipresent *Isotelus gigas*, a species that constitutes one of the area's most renowned, studied, and collected trilobites. The debate rages on.

***ISOTELUS GIGAS* DEKAY, 1824**

Middle Ordovician; Rust Formation, Trenton Group; Walcott/Rust Quarry; Trenton Falls, New York, United States; 11.8 cm

Photo courtesy of the Martin Shugar Collection

(TOP, LEFT) **HYPODICRANOTUS STRIATULUS (WALCOTT, 1875)**

Middle Ordovician; Rust Formation, Trenton Group; Walcott/Rust Quarry; Trenton Falls, New York, United States; 3.5 cm

(TOP, RIGHT) **FLEXICALYMENE SENARIA (CONRAD, 1841)**

Middle Ordovician; Rust Formation, Trenton Group; Walcott/Rust Quarry; Trenton Falls, New York, United States; 4.2 cm

Photo courtesy of the Martin Shugar Collection

(LEFT) **ACHATELLA ACHATES (BILLINGS, 1860)**

Middle Ordovician; Rust Formation, Trenton Group; Walcott/Rust Quarry; Trenton Falls, New York, United States; 3.4 cm

(LEFT) *BUMASTOIDES HOLEI* (FOERSTE, 1920)

Middle Ordovician; Rust Formation, Trenton group; Walcott/Rust Quarry; Trenton Falls, New York, United States; 5.6 cm

(BOTTOM) *SPHAEROCORYPHE ROBUSTA* WALCOTT, 1875

Middle Ordovician; Rust Formation, Trenton group; Walcott/Rust Quarry; Trenton Falls, New York, United States; 1.2 cm

49

10 DESERVING-OF-MENTION TRILOBITES

As the various top 10 sections that comprise *The Trilobite Collector's Guide* were being finalized, an unexpected development began to make its presence known. Despite the intended breadth and scope of this volume's editorial focus, it became abundantly clear that there simply weren't going to be enough categories to squeeze in all the specimens originally planned for inclusion. To remedy this situation, an additional chapter was created to provide sufficient space for some of the trilobite world's most important, interesting, and deserving species. None of these calcite-covered creatures rank among the largest, rarest, oldest, or most colorful examples emerging from their taxonomic class. To the contrary, they're included here expressly because they didn't neatly fit any of this tome's previously prescribed Paleozoic parameters! But these spectacular specimens clearly represent some the most significant trilobites ever to emerge from this noble arthropod line—fossils that unquestionably deserve our respect, recognition, and awe. The photos and descriptions in this chapter have not been chosen to reflect a storied paleontological principle, highlight the contents of a legendary quarry, or reveal a previously mysterious morphological trait. They are being presented here because each trilobite's inherent beauty, comparative rarity, and stunning strangeness make them worthy of being seen, studied, collected, and admired.

Here are 10 trilobites that for varied reasons deserve a prominent place in this book.

KOCHINA VESTITA

No other Paleozoic locality on the North American continent can match the majestic state of Utah in terms of sheer volume of described trilobite material. Hundreds of distinct, scientifically recognized species have emerged from the state's various sedimentary strata, ranging from the incredibly common *Elrathia kingii* to the super rare *Kochina vestita*, a Middle Cambrian trilobite predominantly found in the limestone layers of the high-altitude Spence Shale. In the last six-plus decades, beautifully detailed *Kochina* specimens up to 15 centimeters in length have been retrieved, most featuring a thick black calcite shell. No more than a dozen relatively complete examples of this somewhat tri angular-shaped multisegmented species have ever been unearthed, and a significant percentage of those are molts lacking their free cheeks.

ISOTELUS VIOLAENSIS

Oklahoma is justly renowned for the diversity, preservation, and beauty of its indigenous trilobite fauna. Even a moderately sized fossil collection is virtually guaranteed to feature at least one representative sample drawn from the Sooner State's arthropod-rich Paleozoic landscape. In contrast to the dozens of comparatively common trilobite species that have been recovered by the hundreds over the last half century in the state's bountiful Devonian and Silurian deposits, a few trilo-types remain that are so infrequently seen in complete form that they merit special mention. Near the top of any such arthropod-inspired compendium

ISOTELUS VIOLAENSIS AMATI, 2014
Upper Ordovician; Viola Springs Formation; Pontotoc County, Oklahoma, United States; 14 cm

would be a large species of recently described *Isotelus* that lurks in the rarely explored Ordovician layers of the Viola Springs Limestone, an outcrop that emerges in various spots throughout rural Pontotoc County. Despite years of sporadic searching, less than half a dozen fully articulated examples of this unusual species are known.

REDLICHIA KAIYANGENSIS

Since the early days of this century, China has continually amazed the trilo-centric community with the depth and breadth of its fossiliferous holdings. From

(OPPOSITE PAGE) **KOCHINA VESTITA (RESSER, 1939)**
Middle Cambrian, Miaolingian Series, Wuliuan Stage; Langston Formation, Spence Shale Member; Box Elder and Cache counties, Utah, United States; 12.8 cm

Lower Cambrian to Upper Devonian, this expansive Asian nation has produced a dizzying array of genera that have been unearthed everywhere from the urbanized suburbs of Beijing to the rustic outskirts of Kunming. Among the various trilobites to emerge from this imposing and mysterious land in the last few decades have been a series of previously unseen and unknown *Redlichia* species that have quickly caught the attention of the academic and collecting communities. One of the more visually stunning examples of these recent discoveries would clearly be *Redlichia kaiyangensis*, a Lower Cambrian trilobite with genal spines strangely angled atop its rectangular-shaped cephalon.

OLENOIDES VALI

Throughout the last seven-plus decades, the Gunther family of Lehi, Utah, has played a vital role in the discovery and eventual naming of many unique trilobite species found throughout their native state. The work of the Gunther clan continues unabated today, with new generations of the family exploring many of central Utah's most renowned Cambrian layers, and new species being revealed on a regular basis due to their efforts. In fact, Val Gunther, for whom the rare trilobite *Olenoides vali* is named, served as coauthor of the well-received 2015 volume, *Exceptional Cambrian Fossils from Utah*, which for the first time identified many of the state's rarer trilobite species, including, somewhat ironically, *Olenoides vali*. In addition, Val's son Glade has been honored through the Utah trilobite *Athabaskia gladei*.

BRISTOLIA CF. PARRYI

One of the great pleasures associated with collecting and studying trilobites occurs when a complete example of a supposedly new species is discovered and tentatively identified. In recent years, this has been the scenario surrounding an unusual species of *Bristolia* found in the Lower Cambrian layers of California's Nopah Range. These specimens bear a clear morphological similarity to the scientifically recognized species *B. parryi* (known almost exclusively from disarticulated fragments), and many within the fossil community almost immediately began to arbitrarily assign that name to this "new" trilobite. However, in this case, a "cf." has been added to that classification, meaning that one should "compare" these new specimens to those already assigned to *B. parryi* rather than assuming that it is, in fact, an example of that previously recognized species.

CONOLICHAS SP.

The variety and abundance of trilobite material that has come to light throughout the British Isles in the last millennia is astonishing. From giant Cambrian *Paradoxides* to the comparatively diminutive Silurian *Calymene* found in abundance in the English Midlands, Great Britain seems to possess everything any inveterate trilobite connoisseur could desire. But even among this bounty of well-known and often studied Paleozoic biomatter (trilobite studies first transpired in this part of the world during the middle of the eighteenth century), examples occasionally emerge that in their singularly distinctive rarity and beauty take even the most jaded collector's breath away. Such is unquestionably the case with the magnificent *Conolichas* specimens found in the Ordovician-age mudstone layers of Scotland—a genus upon which additional studies will need to be conducted before a proper species name can be assigned.

TOXOCHASMOPS EXTENSUS

Various species and subspecies of the genus *Chasmops* have long been recognized from key

REDLICHIA KAIYANGENSIS
CHANG AND LIN, 1974
Lower Cambrian, Botomian; Wulongqing
Formation; Guanshan Fauna; Kunming,
China; 9.1 cm

Ordovician fossil repositories located throughout the European continent. These include attractively mineralized examples found in Scotland's renowned Girvan outcrops (where it is known as *Toxochasmops bissetti*), as well as those from western Russia's famed Asery Horizon where these gracefully conformed trilobites have been identified as *Prochasmops praecrurrens*. From nearby Norway, however, complete examples of *Toxochasmops extensus* are exceedingly rare, primarily due to two significant factors. One is that relatively few fossil-seeking souls have ventured to the beautiful Mjosa Lake region where these specimens are extracted. The other is that the hard sedimentary strata in which these trilobites have been preserved for 450 million years has proven particularly difficult to properly prepare.

DOLICHOHARPES DENTONI

Trilobites of the order Harpetida have always attracted their fair share of collector attention due to their irrefutably iconic Paleozoic appearance. Indeed, with a little well-placed imaginative gloss, trilobites such as *Dolichoharpes dentoni* seem to resemble some futuristic design for a sci-fi interstellar vehicle. With its sleek ovate cephalic brim and spurious space-age design, these 450-million-year-old Upper Ordovician specimens drawn from Ontario's renowned Bobcaygeon Formation seem to defy the bounds of time, much to the delight of trilobite enthusiasts everywhere, most of whom yearn to possess even a fractured fragment of this rarely encountered and eminently elegant species.

METOPOLICHAS PLATYRHINUS

In a Paleozoic domain inhabited by countless examples of spectacularly strange species—many that would make even the most ambitious abstract artists tip their chapeau in appreciation—some trilobites are so blatantly bizarre that they defy both human imagination and immediate belief. A prime example of this phenomenon is *Metopolichas platyrhinus*, a 460-million-year-old species culled from the prolific Ordovician outcrops situated near St. Petersburg, Russia. With its elongated body and outlandish duckbill-like cephalic extension, this species indubitably ranks among the most unusual and visually stunning trilobites in the natural world. It is believed by some scientists that this lichid's elongated "nose" was used to stir up seafloor sediments during the trilobite's continual hunt for new food resources.

OLENELLUS HOWELLI

Sometimes it takes a bit of after-hours scientific research to properly classify (or reclassify) even the most well-established and long-accepted trilobite species. For decades, an attractive Lower Cambrian olenellid from the Combined Metals Member of Nevada was recognized as the popular collector's favorite, *Olenellus gilberti.* In the early years of the twenty-first century, a new academic investigation determined that the specimen was more likely a rare example of the closely related *Olenellus howelli.* Such a name-altering academic decision does little to change the inherent beauty or desirability of such a trilobite, but it does affect the average collector in a very fundamental way. After all, they must subsequently generate new identifying labels in their seemingly never-ending quest to remain up to date with these arthropod-inspired adjustments.

(TOP) BRISTOLIA CF. "PARRYI"

Lower Cambrian; Goldace Limestone Formation;
Nopah Range, California, United States; 4.1 cm

**(LEFT) OLENELLUS HOWELLI PALMER,
1998**

Lower Cambrian; Combined Metals Formation;
Chief Mountain, Nevada, United States; 4.4 cm

***TOXOCHASMOPS (TOXOCHASMOPS) EXTENSUS* (BOECK, 1838)**

Upper Ordovician; Furuberget Formation; Mjosa; Lake region, Norway; 5.7 cm

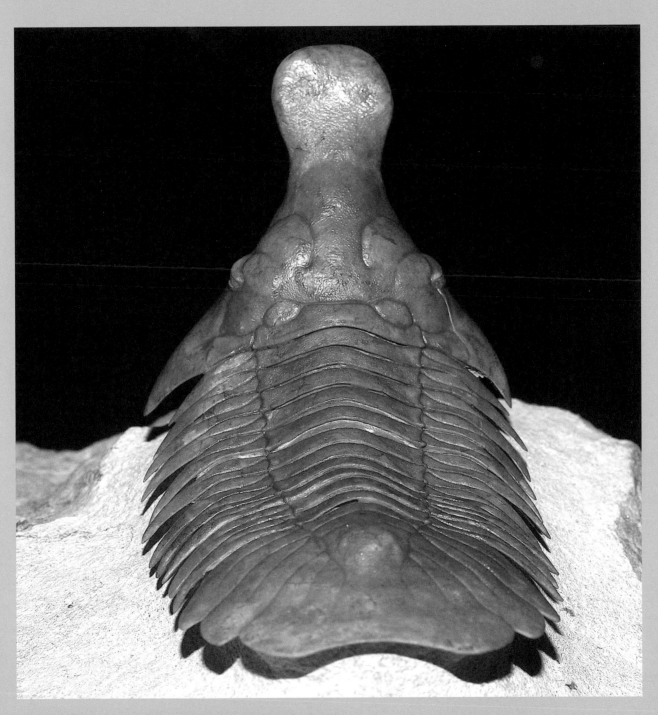

***METOPOLICHAS PLATYRHINUS* (SCHMIDT, 1907)**

Middle Ordovician, Llanvirnian; Kunda Regional Stage (Hunderum-Valaste); Sillarou-Obuchov Formations; Volchov River Valley, St. Petersburg, Russia; 10 cm

***SOKHRITA CF. SOLITARIA* (BARRANDE, 1846) AND CRINOIDS**

Upper Ordovician; Tisserdmine, Erfoud; Morocco; 4.4 cm

10 REVEALING ROCHESTER SHALE TRILOBITES

In the early 1800s construction began on the Erie Canal, the New York State waterway that stretches for nearly 600 kilometers between Albany and Buffalo and connects the Hudson River with Lake Erie. Almost immediately, canal workers operating near the dig's western terminus began to discover strange forms in the 420-million-year-old blocks of Silurian mudstone they were extracting. There were objects they could recognize—brachiopods and corals—and things they couldn't, most notably trilobites. Some fossils uncovered during that landmark project survived their initial uprooting and still reside in the nearby Buffalo Museum of Science. In the 1830s, one of America's most renowned paleontologists, James Hall (later the director of the Museum of Natural History in Albany), began the first extensive scientific excavation of the rich Silurian outcrops of western New York State, particularly a formation known as the Rochester Shale. Accessing the sedimentary exposures created by the canal dig, which was completed in 1825, Hall collected an amazing array of invertebrate fossils from the formation. His discovery of a nearly intact *Arctinurus boltoni* specimen in 1834 eventually led to the publishing of his landmark three-volume treatise, *The Invertebrate Fossils of New York*. With the appearance of that seminal work in 1847, the legend surrounding the Rochester Shale deposit, and its diverse fossil fauna, began to take shape. For the next 150 years, diggers and collectors from throughout the Northeast would frequently wander up to the Erie Canal cuts. They would wait for a canal lock to drain to the required depth, and then they would rapidly extricate a few fossil prizes from the slippery Silurian exposures before the lock again begin to fill. This work was at best an iffy proposition, and recovery of a complete trilobite was a rare occurrence. It is estimated that no more than half a dozen articulated Rochester Shale *Arctinurus* specimens existed prior to 1991when a large commercial fossil-recovery operation began near the town of Middleport. That quarry has subsequently yielded a veritable treasure trove of Silurian fossils, including hundreds of perfectly preserved trilobites.

Here's a look at 10 of the most famed Rochester Shale trilobite species.

ARCTINURUS BOLTONI

Often called the "King of American Trilobites," especially by those Rochester Shale faithful who specialize in the collection, preparation, and eventual sale of these impressive arthropods, there are few fossil enthusiasts in the world who don't covet a complete, well-preserved example of *Arctinurus boltoni*. Growing to sizes of 15 centimeters or more, perhaps no other trilobite in the Paleozoic realm possesses either the visual impact or the scientific gravitas of this renowned lichid species. It is estimated that more than 300 complete *Arctinurus* have been recovered from the Rochester Shale in the last 30-plus years.

DIACALYMENE SP.

The marked faunal diversity and disparity found in the Rochester Shale has caused some academics to present the Sherlockian speculation that certain Silurian trilobite species thrived in the region's oxygen-rich offshore environment, and others did not. How else can one explain the overwhelming quarry presence of *Calymene niagarensis*, which has been unearthed by the hundreds (if not thousands), and the proportional paucity of the closely related *Diacalymene sp.*, of which only four

ARCTINURUS BOLTONI (BIGSBY, 1825)
Lower Silurian; Rochester Shale Formation; Middleport Quarry; Middleport, New York, United States; 13.5 cm

complete examples have yet been revealed. Perhaps in a slightly deeper or more shallow body of the Silurian sea these lopsided proportions might have been totally reversed.

ILLAENOIDES TRILOBA

Illaenoides triloba represents one of the rarer species known from the Rochester Shale outcrops. Perhaps no more than a score of complete examples of this large Illaenid have been found. Judging by the disarticulated fragments that mark certain sedimentary horizons in the Middleport quarry, this species could have grown to sizes in excess of 12 centimeters. It is believed that some smaller, poorly preserved specimens reside in museum collections, with many of those being incorrectly identified as the superficially similar *Bumastus ioxus.* In contrast, *Illaenoides* has a more streamlined shape and notably smaller eyes.

DICRANOPELTIS NEREUS

A story famous among certain fossil folks describes a 10-centimeter-long *Dicranopeltis nereus* found in the Rochester Shale in the late 1990s—a spectacular specimen that tragically met a premature demise due to a badly aimed hammer strike during initial extraction. Despite (or perhaps due to) such discussed disasters, this attractive, pustule-covered lichid remains among the most coveted of all formation faunal elements, and less than two dozen fully articulated examples are known to exist. As a side note, one recognized specimen is only a single centimeter in length, and it is sometimes jocularly referred to as the "*Nanodicrano.*"

CALYMENE NIAGARENSIS

This classic trilobite is perhaps the most prolific species found in the 13-acre confines of the Rochester Shale quarry located in upstate New York. Since commercial digging began at this site in 1991, it is estimated that more than 1,000 examples of *Calymene niagarensis* have been unearthed, occasionally on plates presenting up to a dozen of these 4-centimeter-long arthropods in tightly packed fossilized formation. Some knowledgeable fossil enthusiasts believe that these mass mortality plates may reflect true-life mating or molting assemblages.

TRIMERUS DELPHINOCEPHALUS

These large smooth-shelled homalonotids (that routinely exceed 18 centimeters in length) were initially recognized from the dozens of disarticulated trilobite heads and tails that frequently littered the original nineteenth century Erie Canal cuts. Once more extensive regional explorations began in the early 1990s, however, complete *Trimerus* examples became relatively common. Virtually all of these three-dimensionally preserved specimens have been somewhat compressed on one side of their carapace, a phenomenon that provides their fossilized forms with a slightly lopsided appearance.

BUMASTUS IOXUS

With strikingly similar examples of this popular genus appearing in the Silurian-age outcrops of England, Sweden, and Indiana, a top-quality *Bumastus* remains an essential component of any comprehensive trilobite collection. The 3-centimeter-long *B. ioxus* specimens that emerge from the Rochester Shale are numerous but rarely complete, frequently appearing as molts lacking their free cheeks. Judging by the rarity of this genus in other worldwide Paleozoic outcrops, if hobbyists possess a fully articulated *Bumastus* in their collection, odds are that the trilobite hails from this fossil-filled formation located on the outskirts of Middleport, New York.

DALMANITES LIMULURUS

Once considered a rare element of the Rochester Shale trilobite fauna, especially in fully articulated form, in the last 30-odd years literally hundreds of complete examples of the elegant, 3- to 10-centimeter *Dalmanites limulurus* have been unearthed, occasionally on plates featuring up to 40 overlapping specimens. Upon initial discovery, the turret-like compound eyes of this graceful species often adhere to the negative side of the matrix. It requires delicate work by a skilled preparator to properly extract these ocular outlets and place them in their correct anatomical spot on the trilobite's gently convex cephalon.

DIACALYMENE SP.

Lower Silurian; Rochester Shale Formation; Middleport Quarry;
Middleport, New York, United States; 6.5 cm

CYBANTYX SP.

Despite all the scientific studies that have taken place over the last two centuries involving the Rochester Shale's indigenous fossils, key questions still surround the identity of certain faunal components. Many of these quarry queries focus on the proper taxonomic classification of the formation's 18 recognized trilobite species. One of the most intriguing of these lingering issues revolves around this large (up to 12 centimeter) species, which until recently has commonly been called *Bumastus sp.* Due to a variety of academic assumptions (i.e., the Rochester Shale being part of the Homerian stage of the Silurian Period), it appears that this symmetrically shaped trilobite may, in fact, be a member of the genus *Cybantyx*.

DECOROPROETUS CORYCOEUS

As most Paleozoic enthusiasts are well-aware, proetids were the most successful of all trilobite orders in terms of their longevity. They existed from the middle stages of the Cambrian Period until the demise of the entire trilobite line at the end of the Permian Period, more than 250 million years later. Despite this time-defying success, the proetid presence throughout the Silurian was not particularly notable. Small, compact species such as *Decoroproetus corycoeus* provide telltale signs that explain why these trilobites proved so successful in navigating evolution's often convoluted path. This diminutive species—usually not exceeding 2 centimeters—ranks among the Rochester Shale's scarcest faunal components.

(OPPOSITE PAGE) **ILLAENOIDES CF. TRILOBA (WELLER, 1907)**

Lower Silurian; Rochester Shale Formation; Middleport Quarry;
Middleport, New York, United States; 10.8 cm

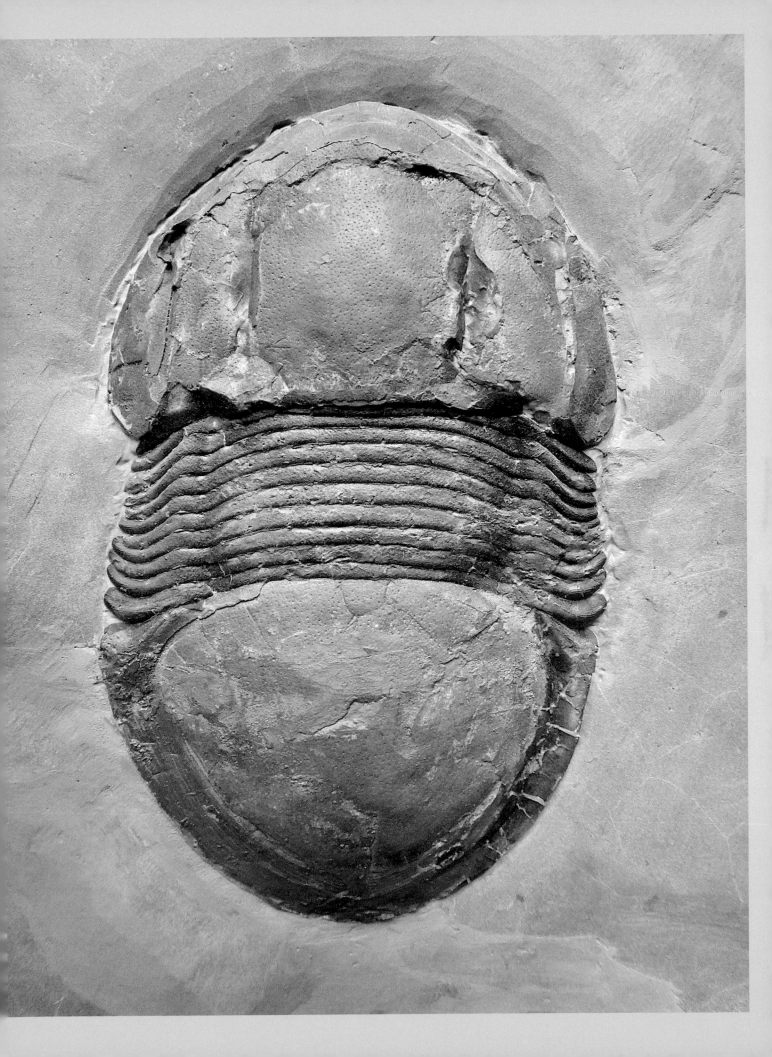

(LEFT) DICRANOPELTIS NEREUS (HALL, 1863)

Lower Silurian; Rochester Shale Formation; Middleport Quarry; Middleport, New York, United States; 6.8 cm

(BOTTOM) CALYMENE NIAGARENSIS (HALL, 1852)

Lower Silurian; Rochester Shale Formation; Middleport Quarry; Middleport, New York, United States; largest trilobite 3.8 cm

BUMASTUS IOXUS HALL, 1868

Lower Silurian; Rochester Shale Formation; Middleport Quarry; Middleport, New York, United States; largest trilobite 2.8 cm

DALMANITES LIMULURUS (GREEN, 1832)

Lower Silurian; Rochester Shale Formation; Middleport Quarry; Middleport, New York, United States; largest trilobite 6.2 cm

51

10 TANTALIZING TRILOBITE TEASERS

M ost trilobite collectors focus their energies primarily on the acquisition and subsequent display of complete specimens. Partial or poorly preserved examples are often discarded by those who venture into the wilderness in pursuit of these long gone Paleozoic relics. Any pleasure derived from callous-causing hours spent breaking rock in a tree-lined quarry or a stream-abutting sedimentary outcrop can quickly vanish when, after a little prep, it's discovered that your hoped for trilobite trophy is little more than a disarticulated fossil fragment. However, trilobite enthusiasts shouldn't become overly distraught when confronted with such an unexpected and unwanted development. A singularly distinctive remnant of Earth's distant past may still be gleaned from such an anticlimactic experience. Whether it is the isolated cephalon of a super rare *Ceratocephala* from the Silurian of England or the pygidium of a hitherto undescribed *Arctinurus* found amid the cliffside scree of Indiana, these calcite-coated bits of trilobite anatomy can provide both scientists and collectors with invaluable insight into species that are virtually or totally unknown in complete form. Unquestionably, these tantalizing bits and pieces of Paleozoic detritus may at first be viewed as nothing more than spirit-crushing disappointments by those who discover them. But once that initial dismay begins to subside, these incomplete specimens can be better viewed in their proper paleontological context. Indeed, they provide unique and often compelling evidence of the incredible morphological diversity that the trilobite class enjoyed during its quarter-billion-year march through the Deep Time seas.

Here's a salute to 10 tantalizing trilobite teasers.

BOWMANIA AMERICANA

Finding even scattered pieces of this extraordinary species amid the rugged Upper Cambrian outcrops of the American West can be cause for pure paleontological celebration. With its highly distinctive ring of cephalon-encircling spikes and long, gently curving genal spines—which some imaginative fossil enthusiasts believe resemble a stereotypical Native American headdress—few trilobites in the natural world appear as unusual as *Bowmania americana*, and even fewer are as

BOWMANIA AMERICANA (WALCOTT, 1884)

Upper Cambrian; White Pine County; Nevada, United States; 5 cm

This is a ventral cephalon.

coveted by collectors around the globe. In fact, the negligible number of partial examples that have emerged from secretive limestone layers in central Nevada have been uniformly preserved in a ventral state, thus making many of the morphological details associated with this elusive trilo-type speculative at best.

BRISTOLIA ANTEROS

Although always rare, various species of the Lower Cambrian genus *Bristolia* continue to emerge from properly aged sedimentary outcroppings throughout the western United States. Many of these notable discoveries rank among the most eagerly collected examples in the entire trilobite domain. With their distinctively flared genal spines and graceful morphological profile, it's easy to understand the wide-ranging appeal of these early representatives of the ancient arthropod line. But even among trilobites drawn from such a familiar and well-established genus, species such as *Bristolia anteros* remain unknown in complete form to both scientific and collecting communities.

ARCTINURUS SP.

There is no question that *Arctinurus* has achieved a richly deserved status among the most instantly recognized and best-known of all trilobite genera. Since its initial discovery in the Rochester Shale in the early 1800s, scores of complete examples of this impressive lichid have been uncovered in the thickly banded limestone layers of that prolific western New York site. In other similarly aged Silurian localities, including the Jupiter Formation of eastern Quebec and the Waldron Shale of southern Indiana (an outcrop perhaps best known for its large, complete *Glyptambon* specimens), only frustratingly disarticulated *Arctinurus* fragments have been unearthed, despite dozens of these partial pieces—usually well-preserved pygidia—being found. Such Waldron fragments provide ample proof, however, that this remarkable species may have once been prevalent throughout the seas that covered this famed midwestern locale 425 million years ago.

NEOPROBOLIUM OKLAHOMAE

Much like Sasquatch or the Loch Ness Monster, certain trilobite species remain squarely sequestered in the realm of myth and legend. Tantalizing traces of their possible existence occasionally emerge to further encourage the true believers that pursue them. But too often those evidentiary bits prove false, or at best are misleading. This has apparently been the long-standing situation surrounding the Devonian trilobite *Neoprobolium oklahomae*, a dalmantid species possessing a forked rostral extension that has been independently found (often attached to a disarticulated cephalon) in the fossiliferous outcrops of southern Oklahoma. Despite this comparatively compelling string of clues, no complete example that would unequivocally confirm the existence of this controversial species has yet been revealed.

CERATOCEPHALA BARRANDEI

There are probably less than a handful of complete examples of the spinose Silurian trilobite *Ceratocephala barrandei* currently residing in the centuries-old repositories provided by leading British natural history museums. Even the most dedicated (and ambitious) hobbyists should perhaps limit their collecting aspirations to obtaining no more than a mere cephalon of this hard to find species—a head that has long been separated from the fossilized remains of this midsized odontopleurid. The most prominent *Ceratocephala* species hail from the Much Wenlock Limestone outcrops that

BRISTOLIA ANTEROS (PALMER IN PALMER AND HALLEY, 1979)

Lower Cambrian; Carrara Formation; Amargosa Valley; Nye County, Nevada, United States; 3.4 cm across genal spines

comprise the legendary Wren's Nest site near the town of Dudley, England. In recent years, similarly aged pockets skirting the Midlands village of Malvern have also yielded well-preserved fragments of this elusive bug. A similar species can be found in closely aligned Czech formations.

TERATASPIS GRANDIS

Perhaps no partial trilobite remains are as renowned, revered, or cherished as those attached to—or detached from—the fabled Devonian species *Terataspis grandis*. For decades dating back to the early years of the twentieth century, mysterious pustule-covered pieces of this over-sized lichid (which when complete could have grown to lengths exceeding 35 centimeters) have been unearthed in sedimentary deposits throughout western New York State and southern Ontario. Rumors of complete specimens abound but, except for one documented dolomitic example currently on display in Toronto's Royal Ontario Museum, the existence of these iconic Paleozoic prizes have remained unverified.

NEODREPANURA PREMESNILI

Many years ago, while casually strolling through New York's famed Chinatown, a noted trilobite enthusiast encountered a curio shop selling dozens of small "swallow stone" blocks covered in scattered bits of trilobite debris, each carefully perched atop a shiny wooden base. This collector later discovered that these so-called swallows were, in fact, the disarticulated pygidia of diminutive Cambrian trilobites, including *Neodrepanura premesnili*, examples of which have been noted and revered in certain segments of Chinese society for nearly a thousand years. However, despite the overwhelming volume of trilobite parts uncovered in Shandong Province's Kushan Formation in the ensuing centuries, it is believed that the first complete example of *Neodrepaneura* was only found during the second decade of the twenty-first century. Since then perhaps a half-dozen articulated specimens have emerged.

CHEIRURUS SP.

It's interesting to listen as those who revel in field collecting tell their (often tall) tales about certain trilobite species that have eluded their fossil-finding grasp. Much like deep sea fishermen who love to speak of "the one that got away," trilobite diggers seem to enjoy regaling any interested parties with stories of specific trilo-types that have left behind tantalizing clues of their long-ago lives but have continually proven highly evasive as fully articulated examples. One such trilobite is a species of undescribed cheirurid that lurks in Nevada's Ordovician-age Antelope Valley Limestone. That difficult to reach repository is perhaps most noted for producing magnificent complete specimens of the highly collectible trilobite *Pseudomera barrandei*. In the case of Antelope Valley's

faunal cheirurid, however, the search for a complete example continues.

FENESTRASPIS AMAUTA

In the concretion-laced Devonian deposits that sit high atop Bolivia's soaring Altiplano, the trilobite fragments found are frequently so bizarre that they prove difficult for the human mind to fully fathom. Some of these brain-bending trilo-bits derive from the rarely encountered phacopid *Fenestraspis amauta*, a species that features the largest compound eyes in the entire trilobite world. With its dense, lens-covered optic columns occasionally reaching heights of more than 3 centimeters, these singularly distinctive ocular outlets (which, unfortunately, are often discovered alone and apart from the surrounding trilobite) offer mesmerizing hints as to what a complete example of *Fenestraspis* may have looked like.

BURMEISTERELLA ACULEATA

Many of the trilobites emerging from the thinly bedded charcoal-gray siltstone sheets that distinguish Germany's Hunsruck Slate are relatively common. Due to decades of quarrying conducted in and around this world-renowned Devonian site—primarily to provide the local constituency with roofing and paving tiles—scores of complete specimens have been uncovered, most representing the familiar phacopid *Chotecops ferdinandi*. But mixed among these somewhat expected finds (which also include an array of magnificently preserved starfish and crinoids) are amazing partial trilobite pieces that on rare occasions may include the disarticulated detritus of the distinctive homalonotid *Burmeisterella aculeata*. Although a handful of discoveries have come close, no complete example of this large (8- to 12-centimeter) trilobite has yet to emerge from Hunsruck's legendary layers.

(TOP) *CERATOCEPHALA BARRANDEI* (SALTER, 1853)

Middle Silurian, Wenlock Series, Homerian Stage; Coalbrookdale Formation; Storridge, Worcestershire, England; 3 cm across genal spines

(LEFT) A SO-CALLED SWALLOW STONE FROM CHINA THAT FEATURES DISARTICULATED CAMBRIAN TRILOBITE PYGIDIA

NEODREPANURA PREMESNILI (BERGERON, 1899)

Middle Cambrian; Kushan Formation; Shandong Province, China; 3.2 cm

ARAPAHOIA SP.

Middle Cambrian; Weeks Formation; House Range; Utah, United States; 1.2 cm

52

10 DISTINGUISHED DUDLEY TRILOBITES

The Silurian-age outcrops in and around Dudley, England, have been exhumed, examined, and collected for nearly a millennium. During that time, the paleontological prominence of Dudley's Wren's Nest location has continually grown in esteem among trilobite enthusiasts. Indeed, the vicinity's Much Wenlock deposits now rank among the most renowned fossil repositories in all of Europe and are among the most prominent Silurian strata in the entire world. Quarrying the area's thick limestone layers was initially undertaken to acquire building materials for some of the town's most ambitious architectural projects. In fact, many of Dudley's famous landmarks have been constructed solely out of the 420-million-year-old sedimentary blocks brought forth from the nearby hillsides. Understandably, those who first explored these horizons 900 years ago were both confused and confounded by the numerous fossils they encountered, especially the beautifully preserved trilobites. To them, these strange "insects made of stone" must have appeared to be inexplicable relics from some distant and mysterious time. Soon after the dawn of the eighteenth century, however, a few daring and enterprising souls began to more closely investigate these rugged outcroppings and to appreciate Dudley's wide variety of fossilized life-forms—especially the scores of calcite-coated trilobite species that they produced. In the following centuries, interest in the natural sciences began to slowly expand in England's upper classes, and collecting fossils throughout the Wren's Nest vicinity became something of a cottage industry. By the mid-1800s, many locals would scour the site for interesting trilobite

DALMANITES MYOPS (KÖNIG, 1825)
Middle Silurian, Wenlock Series, Homerian Stage; Much Wenlock
Limestone Formation, Upper quarried limestone; Wren's Nest
Hill, Dudley, West Midlands, England; 7.1 cm

material, then set up small roadside stands offering choice Paleozoic morsels to the surprisingly large hordes of well-to-do weekend visitors, virtually all of whom were intent on acquiring their share of Silurian-age souvenirs.

Here's a look at 10 of Dudley's most renowned trilobite species.

DALMANITES MYOPS

Due to the apparently gentle nature of the ocean currents that enveloped the Wren's Nest environs during the Silurian, Dudley's future fossils were afforded the luxury of resting in a relatively undisturbed state as they slowly became covered by layer upon layer of mineral-laden sea silt. Over the ensuing eons, these sediments were gradually transformed into well-defined limestone sheets by the internal forces housed in planet Earth. Despite the incredible compression and shearing torque that regularly accompanies the fossilization process, the trilobites of Dudley—such as the gracefully shaped *Dalmanites myops* that ranks among the formation's more prolific species—have become world-renowned for their amazingly lifelike, three-dimensional preservation.

DEIPHON BARRANDEI

Among the tens of thousands of described trilobite species that have emerged from Earth's sedimentary layers over the centuries, a few have achieved an almost mythical status among properly schooled enthusiasts and academics. Whether due to their strange shape, exceptional preservation, or unconventional place of discovery, collectors' dreams often feature these coveted specimens. Near the apex of any such consideration lurks *Deiphon barrandei*, a small (2- to 4-centimeter) bubble-nosed cheirurid that represents one of Dudley's rarest and most eagerly sought faunal elements. Indeed, the hard to obtain fossilized remnants of this bizarre species invariably represent the crown jewels of any trilobite collection, either public or private.

STAUROCEPHALUS MURCHISONI

A surprising number of species that existed during the trilobites' quarter-billion-year crawl through Deep Time exhibit a most unusual morphological feature—a pronounced "ball" perched at the front of their glabella. What possible evolutionary advantage this prominent clownlike "nose" provided has been speculated upon by the scientific community ever since this strange anatomical adornment was first noted in the fossil record. Lurking in Dudley's abundant fauna is perhaps the most notable example of this unusual phenomenon, the rare, diminutive (under 2 centimeters long), but highly distinctive *Staurocephalus murchisoni*—named in honor of the famed Scottish geologist Sir Roderick Murchison—which vividly demonstrates the trilobite class's bulbous cephalic proclivities.

EOPHACOPS MUSHENI

It is somewhat surprising that many of the Wren's Nest trilobite examples found in the nineteenth century—when the concept of fossil preparation usually involved hand-wielding anything from small pocket knives and spikelike nails to rather cumbersome dental tools—still hold their own in terms of quality, rarity, and desirability. In fact, the trilobites unearthed during those early digs, certainly including the relatively common species *Eophacops musheni*, retain a special cachet among hobbyists. Some seek these so-called Victorian Collection pieces and choose them over more recently discovered and air abrasively prepared Dudley examples.

CYBANTYX ANGLYPTOS

For many decades, if not centuries, this rare medium-sized trilobite found in Dudley's prolific fossil-bearing formations was recognized as *Bumastus barriensis*, a species strikingly similar to one long known from the slightly older Silurian rocks of nearby Shropshire. As increased scientific research was undertaken, academic thought shifted toward believing that the three-dimensionally preserved genus being discovered in the Wren Nest site's hard limestone layers was, in truth, *Cybantyx anglyptos*. Both genera appeared nearly

DEIPHON BARRANDEI (WHITTARD, 1934)
Middle Silurian, Wenlockian Stage, Homerian Substage; Much Wenlock Limestone Formation; Wren's Nest Hill, Dudley, West Midlands, England; 3.2 cm

identical. It is the size and placement of each trilobite's eyes as well as the indicative Silurian time stage during which they lived (Homerian vs. Sheinwoodian) that supplied the keys to unlocking this ongoing and still somewhat controversial academic investigation.

SPHAEREXOCHUS MIRUS

Due to the extensive excavations that have taken place throughout the centuries in and around the Wren's Nest site, few other paleontological pockets in the world—including such equally bountiful Silurian repositories as New York State's Rochester Shale and Sweden's Hemse Formation—have had their trilobite fauna more thoroughly recognized,

CYBANTYX ANAGLYPTOS LANE AND THOMAS, 1978

Middle Silurian, Upper Wenlockian Stage; Coalbrookdale Formation; Malvern, England; 2.7 cm

scrutinized, and analyzed. Perhaps equally as significant to collectors as this documented diversity of Dudley trilobite species is the magnificent manner in which these ancient organisms were preserved once their carapaces sank into the lime-rich mud that then covered the bottom of the area's marine basin. That is certainly true for spectacularly fossilized trilobites such as the small,

elongated cheirurid *Sphaerexochus mirus,* a species that rarely exceeded 4 centimeters in length.

CALYMENE BLUMENBACHII

Today just about any major natural history museum or well-curated private collection will feature an assortment of trilobite specimens derived from the famed fossil-filled Much Wenlock Limestone outcrops that emerge in the Wren's Nest locale. The pervasive 4- to 7-centimeter "Dudley Locust," *Calymene blumenbachii*, is still the most famous—and most common—species found. But even that ubiquitous trilo-type, of which hundreds of complete examples have been recovered throughout the last millennium, has become a rare commodity in the twenty-first century as increasing restrictions now make anything more than surface collecting at the Dudley site (hammers forbidden) a distinct no-no.

ENCRINURUS PUNCTATUS

Over the centuries, complete examples of the rare "strawberry headed" trilobite *Encrinurus punctatus* have been discovered throughout the Wren's Nest locality in sizes ranging from 3 to 6 centimeters. More significant, these specimens frequently demonstrate the resplendent preservation for which Dudley fossils are justifiably renowned. As with most Much Wenlock trilobites, their original exoskeleton has been replicated in a beautiful chocolate-hued calcite that allows for morphological features—including their dramatic eyes and delicate spines—to be captured in minute detail. Once skilled preparation work is done to remove what remains of their surrounding hard rock matrix, these ancient sea creatures appear little the worse for wear from their 400-odd-million-year journey through time.

TRIMERUS DELPHINOCEPHALUS

You have encountered *Trimerus delphinocephalus* in a variety of fossiliferous fashions throughout this book. Indeed, with that genus hailing from disparate Silurian outcrops in New York, Sweden, and Indiana, as well as in England, *Trimerus* rates among the planet's most widely dispersed trilobites. But the first time the academic community encountered evidence of this large, smooth-shelled homalonotid was probably an example drawn from the bountiful sedimentary outcrops that characterize Dudley's Wren's Nest locale. Most often such fresh from the field specimens were not complete, usually featuring only a molted head or tail that had become separated from the creature's carapace. However, fully articulated *Trimerus*, some up to 16 centimeters in length, have been revealed on rare occasions, much to the delight of anyone fortunate enough to have uncovered such an impressive relic of our planet's distant past.

KTENOURA RETROSPINOSA

Wren's Nest trilobites, such as the intricately constructed and exceedingly rare chcirurid *Ktenoura retrospinosa*, remain among the most admired and desired Paleozoic relics in the world. That's especially true for fossil connoisseurs who revel in both the history and the allure exhibited by these magnificent examples of nature's primal handiwork. Even two centuries after visitors initially began exploring the area's rich Silurian outcroppings with increased academic vigor, Dudley proudly retains its title as one of the planet's premier paleontological sites. That is a distinction it will not relinquish any time soon.

(TOP) *SPHAEREXOCHUS (SPHAEREXOCHUS) MIRUS* BEYRICH, 1845

Middle Silurian, Upper Wenlockian Stage, Homerian Substage; Much Wenlock Limestone Formation; Wrens Nest Hill, Dudley, West Midlands, England; 3 cm

(LEFT) *ENCRINURUS PUNCTATUS* (BRUENNICH, 1781)

Middle Silurian, Upper Wenlockian Stage, Homerian Substage; Much Wenlock Limestone Formation; Wren's Nest Hill, Dudley, West Midlands, England; 4.5 cm

(OPPOSITE PAGE) *TRIMERUS DELPHINOCEPHALUS* (GREEN, 1832)

Middle Silurian, Upper Wenlockian Stage, Homerian Substage; Much Wenlock Limestone Formation; Wren's Nest Hill, Dudley, West Midlands, England; 13.6 cm

Photo courtesy of the Martin Shugar Collection

***KTENOURA RETROSPINOSA* LANE, 1971**

Middle Silurian, Upper Wenlockian Stage, Homerian Substage; Much Wenlock Limestone Formation; Wren's Nest Hill, Dudley, West Midlands, England; 2.7 cm

Photo courtesy of the Martin Shugar Collection

Final Thought
(Inspired by Carl Sagan)

The universe is 13.7 billion years old. If that incomprehensible span of time was somehow compressed into a single year starting on January 1, our Milky Way Galaxy would initially coalesce on May 2. Earth would form on September 15. The earliest single-celled organisms would appear in the planet's oceans on December 13. The entire 270-million-year reign of trilobites would last for one week, from December 18 until December 25. Dinosaurs would first rumble across the global landscape on the afternoon of December 27 and be gone by midmorning on December 30. The earliest human ancestors would emerge at 10:30 P.M. on the evening of December 31, with the first *Homo sapiens* appearing at 11:48 P.M. on that same New Year's Eve.

KIERARGES MORRISONI CORBACHO 2014

Lower Ordovician, Arenigian; Upper Fezouata Shales; Zagora, Morocco; 12.4 cm

SPECIAL THANKS

Throughout my 30-plus years of collecting trilobites, people have continually stepped forward to further inspire my peculiar Paleozoic passion. These individuals—with whom I have either traded specimens, swapped gossip, garnered knowledge, or shared insight—include fellow trilobite enthusiasts Martin Shugar, Jodi Summers, Sam Stubbs, Ray Meyer, Bill Barker, Warren Getler, Val Gunther, Glade Gunther, Perry and Maria Damiani, Phil Isotalo, Doug DeRosear, Riccardo Levi-Setti, Jack Shirley, Kirk Johnson, Robert Schacht, Kevin Brett, Terry Abbott, Richie Kurkewicz, Carles Coll, Greg Heimlich, Matt Phillips, Maximo Rojo, Eugene Thomas, James Cook, Carlo Kier, Brian Whiteley, Tom Whiteley, Jason Cooper, Dan Cooper, Mark Norell, Tom Lindgren, Niles Eldredge, Markus Martin, George Lee, Robert Kennedy, Tom Johnson, Shengpeng Li, George Stone, Russ Jacobson, Pierre Marie-Guy, George Ast, and Jake Skabelund. Also, the specimens featured so prominently in *The Trilobite Collector's Guide* would not appear anywhere near as visually appealing without the efforts of talented prep masters Ben Cooper, Alf Cawthorn, Dave Comfort, Scott Vergiels, and Zarko Ljuboja—all of whom manage the seemingly impossible task of transforming slabs of raw trilobite-bearing sedimentary stone into museum-quality showpieces.

Glossary
Trilobite Terminology

In the series of top 10 lists that fill this trilobite-themed book, some terms may be unfamiliar to readers who are not dedicated arthropod aficionados. This abbreviated glossary of occasionally tongue-twisting trilo-terms is provided to enhance your enjoyment of the many trilobites you will encounter in these pages.

Appendages: The antennae, limbs, and walking legs that all trilobites possessed and that on rare occasions appear as fossilized remnants.

Arthropod: The phylum that includes trilobites. All arthropods possess bilateral symmetry, segmented bodies, and chitinous exoskeletons.

Axis: The central lobe of a trilobite's three-lobed body design.

Benthic: Relating to organisms (including many species of trilobites) that live at the bottom of a body of water.

Bug: A "pet name" used by trilobite enthusiasts when they refer to their favorite fossil arthropod.

Cambrian: The 56-million-year-long period during which trilobites first emerged and attained their greatest global diversity.

Carboniferous: Often divided into the Mississippian and Pennsylvanian in North American paleontological parlance, this 60-million-year-long period saw a sharp drop in trilobite speciation.

Cephalon: The head of a trilobite housing the eyes and other important morphological features.

Class: The taxonomic group Trilobita represents the class within which all trilobite orders, families, genera, and species are placed.

ASAPHUS NIESZKOWSKII SCHMIDT, 1898
Upper Ordovician, Caradocian; Kukruse Regional Stage;
Viivikonna Formation; Kingisepp district, St. Petersburg region,
Russia; 1.7 cm

Compound eyes: The multilensed eyes that often are the most pronounced feature of a trilobite cephalon.

Cruziana: Trace fossils comprised primarily of trackways that indicate the walking or crawling lifestyle of a trilobite.

Devonian: The 60-million-year-long period during which trilobite eyes reached their apex of evolutionary design.

Dorsal: The top (back) side of a trilobite.

Doublure: A cephalic or pygidial rim that stretches under the ventral side of the trilobite.

Ediacaran: The 96-million-year-long geologic period that preceded the Cambrian. It was during this time that possible precursors to trilobites first emerged.

Enrollment: A feature that allowed most trilobite species to assume a ball-like shape to best protect their vulnerable ventral underside.

Exoskeleton: The trilobite's hard calcium-covered outer shell.

Fauna: Animals that inhabit a certain region or geological period.

Flora: Plant life that inhabits a certain region or geological period.

Free cheeks: The parts of a trilobite's cephalon surrounding the glabella that can be jettisoned during molting.

Genal spine: A spine emanating from the cheek area of the cephalon.

Genus: The taxonomic group that ranks above species and below family: *Dalmanites limulurus* presents the trilobite's genus followed by its species.

Glabella: The midsection of the trilobite cephalon; it is often convex in shape.

Holochroal: A type of trilobite compound eye in which all of the lenses are in direct contact and are covered by a single calcite surface.

Hydrodynamic: The streamlined shape of some trilobite carapaces that apparently aided them while moving through primeval seas.

Hypostome: The trilobite's mouth plate, which is located on the ventral side of the cephalon.

Isopygous: When the trilobite's head and tail are roughly the same size.

Lagerstatte: A location where the fossil content and preservation is of unique scientific importance.

Mesozoic: The Age of Dinosaurs, the 186-million-year-long era directly following the Paleozoic.

Molting: The action through which a trilobite shed its hard outer shell as an essential part of the growth process.

Negative: The reverse side of a trilobite fossil; it shows a cast of the positive side of the specimen but retains none—or a minimal amount—of the original calcite shell material.

Nonbiomineralized: Lacking the calcite-covering that apparently first emerged with trilobites in the Lower Cambrian, such as the soft-bodied arthropods found in the Burgess Shale.

Opisthothorax: A long, wormlike extension of the thorax that appears on certain Lower Cambrian trilobites and may provide evidence of the trilobites' wormlike predecessors.

Ordovician: The 41-million-year-long period during which some of the largest and strangest trilobite species emerged.

Paleozoic: The 290-million-year-long era that stretched from the Lower Cambrian through the Permian and neatly bookends the rise and eventual fall of trilobites. Paleozoic means "the time of ancient life."

Pelagic: Relating to organisms (including some species of trilobites) that lived amid the upper layers of a body of water.

Permian: At the end of this 47-million-year-long period, the trilobites met their demise, along with 90 percent of the life-forms on Earth.

Pleural lobes: The two lobes that flank the central axial lobe; together the three create a "tri-lobe-ite."

Prosopon: The finely detailed terracing or structures that can appear on the trilobite's outer shell.

Pygidium: The tail end of a trilobite.

Schizochroal: A compound eye featuring separate and distinct lenses.

Silurian: The 25-million-year-long period during which predators such as jawed fish began to further erode the trilobites' marine dominance.

Species: A group of similar organisms that share common attributes: *Olenellus clarki* presents the trilobite's genus followed by its species.

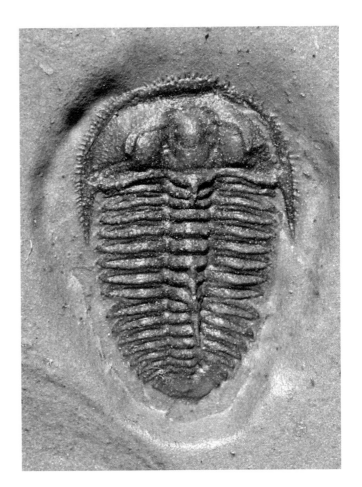

***BOLASPIDELLA DRUMENSIS* ROBISON, 1964**
Middle Cambrian; Pierson Cove Formation; Drum Mountains, Utah, United States; 1.5 cm

Spinose: When prominent spines appear on the trilobite exoskeleton.

Suture: Natural breaks in the trilobite carapace that facilitated the molting process.

Telson: The pronounced tail spine emanating from the pygidium of certain trilobite species.

Thorax: The middle part of the trilobite's anatomy, often featuring between 8 and 15 flexible segments.

Ventral: The underside of a trilobite where fossilized evidence of muscle attachment scars (and occasional soft-body parts) may be seen.

(LEFT) *DICTYELLA LONGISPINA* ZHOU, 1977

Upper Cambrian; Sandu Formation; Jingxi, Guangxi Province, China; 1.6 cm

(BOTTOM, LEFT) *THYSANOPELTIS SPECIOSA* HAWLE AND CORDA, 1847

Middle Devonian, Eifelian; Fezna "white layer"; Jorf, Morocco; 9.8 cm

(BOTTOM, RIGHT) *HOPLOLICHAS AFF PRAEPLAUTINI*

Middle Ordovician, Asery Stage; Gostilitsy Village; St. Petersburg, Russia; 5.1 cm

Index